Springer Series in Materials Science

Volume 305

The Springer Series in Materials Science covers the complete spectrum of materials research and technology, including fundamental principles, physical properties, materials theory and design. Recognizing the increasing importance of materials science in future device technologies, the book titles in this series reflect the state-of-the-art in understanding and controlling the structure and properties of all important classes of materials.

More information about this series at http://www.springer.com/series/856

Yoichi Okimoto · Tomohiko Saitoh ·
Yoshihiko Kobayashi · Sumio Ishihara
Editors

Spin-Crossover Cobaltite

Review and Outlook

Springer

Editors
Yoichi Okimoto
Department of Chemistry
Tokyo Institute of Technology
Tokyo, Japan

Yoshihiko Kobayashi
Department of Physics
Tokyo Medical University
Tokyo, Japan

Tomohiko Saitoh
Department of Applied Physics
Tokyo University of Science
Tokyo, Japan

Sumio Ishihara (Deceased)
Department of Physics
Tohoku University
Sendai, Japan

ISSN 0933-033X ISSN 2196-2812 (electronic)
Springer Series in Materials Science
ISBN 978-981-15-7928-8 ISBN 978-981-15-7929-5 (eBook)
https://doi.org/10.1007/978-981-15-7929-5

This Springer imprint is published by the registered company Springer Nature Singapore Pte Ltd.
The registered company address is: 152 Beach Road, #21-01/04 Gateway East, Singapore 189721, Singapore

In Memory of Sumio Ishihara

Just before the completion of this book, Prof. Sumio Ishihara, one of our fellow editors, passed away at the age of 56. We were greatly shocked to hear of his sudden, untimely passing, and we deeply regret that we cannot show him the book in its published form.

Sumio Ishihara was an active theoretical physicist in condensed matter physics, especially in the field of strongly correlated physics. He dealt with many strongly correlated materials including high-T_c cuprates. In the light of "orbital degrees of freedom" in transition metals as well as on the basis of his own intuition and original ideas, he modeled effective Hamiltonians of those materials and succeeded in describing their electric, magnetic, and optical properties. Throughout his activities as a theoretical physicist, he kept showing us many important aspects of correlated materials using his smart technique of numerical calculations.

He also saw spin-crossover cobaltites, the subject of this book, as a fertile arena for theoretical physics. He predicted and revealed several interesting aspects of cobaltites not only in the ground state but in the excited state as well. We believe that in Chap. 3, the reader can see and find delight in some of his recent works concerning cobaltites, derived from his pioneering ideas and numerical techniques.

He was mild-mannered and liked by everyone. We were always impressed by the sincerity and curiosity shown in his eyes when he engaged in discussions with his colleagues, coworkers, and friends. At the initial stage when this book was proposed to the publisher, there were many problems, but the encouragement we received through the calm but fervent look in his eyes enabled us to go ahead. We are sincerely grateful to him for his efforts in making this book possible, as we pray for his soul on behalf of all the authors of the book.

In dedicating this book to Sumio Ishihara,

Yoichi Okimoto
Tomohiko Saitoh
Yoshihiko Kobayashi

Preface

This book on spin-crossover phenomena of cobalt perovskites was edited to serve as a guide for beginner researchers and/or graduate students as well as a good reference that can give many active researchers a clear view of the properties of cobalt oxides with the historical background of hitherto studies.

Since the condensed matter physics based on the quantum mechanics became a research topic a long time ago, oxide materials containing transition-metal elements have always offered us an important research arena that includes various intriguing electric and magnetic properties, especially in oxide ceramics with 3*d* electrons. Studies were accelerated by the discovery of the copper oxides showing high-Tc superconductivity. The research exploring the novel oxide superconductivity since 1980s has not only reconfirmed the important role of the Coulomb correlation between the 3d electrons first proposed by N. F. Mott but also stimulated the research of various transition-metal oxides other than cuprate, leading to the development of several experimental techniques.

Among the many transition-metal oxides, cobalt perovskites have achieved a unique position in the research field of the oxide materials including the high-Tc superconductivity. The most interesting and widely noticed electronic feature in the cobalt system is spin-crossover phenomena, especially in lanthanum cobalt oxide ($LaCoO_3$). There is a temperature variation of the spin state observed in the transition metal ion with 6 d electrons (e.g., Fe^{2+} or Co^{3+}) in a strong ligand field. In fact, the spin-crossover phenomena have also been found in some classes of iron complexes, as introduced in the Springer chemistry series (Spin crossover in transition metal compound I-III, edited by Gütlich *et al.*). To understand the phenomena, the Tanabe-Sugano diagram plays an essential role on the basis of the energy competition between the crystalline field and the electron correlation factors, e.g., Racah parameters, and well explains the iron complexes even for the photoexcited state. However, such a classical framework is not necessarily relevant in cobalt perovskites. As this book mentions in detail, the discrepancy between the observed experimental results and the Tanabe-Sugano diagram was first observed in 1970s and has since remained unraveled, causing active discussions even now. This is the first reason we have edited a book concerning the spin-crossover cobalt oxide. In the first chapter, we describe the present situation of the study of the spin crossover beyond the Tanabe-Sugano framework, overviewing the history of the research on $LaCoO_3$ during the past quarter of a century.

The second reason is the recent remarkable developments in the study of spin-crossover cobalt perovskites driven by the abovementioned developments. The second chapter describes some new cobaltites, such as layered perovskites and thin films, and details various spin-crossover phenomena. In addition to the simple spin state transition observed in $LaCoO_3$, we further discuss how the spin crossover affects the electronic structure including the lattice structure, electric conductivity, and optical properties.

Another important aspect of the spin-crossover system is photoinduced phase transition, which has been also a famous effect in iron complexes. A recent development of femtosecond laser techniques has enabled us to observe exotic photoinduced phenomena which are different from those in the iron complexes. There are promising developments toward an ultrafast communication and a discovery of new material phases. In this book, we introduce some recent observations of the photoexcited state in some cobalt perovskites. Motivated by the ultrafast experiments, recently, researchers have conducted trials to understand microscopically the photoexcited state on the time scale of femtosecond, in light of the theoretical approach. Theoretical treatments have also been reported for spin-crossover cobalt oxides, and this chapter introduces recent numerical calculations to overview not only the ground state but also the photoexcited state in terms of the developed numerical techniques.

The third reason is the possible applications of the spin-crossover cobaltites. During the past decade, some cobalt oxides have been found to show novel electronic features, such as ionic conductivity, gigantic spontaneous polarization, and thermoelectronics. In the third chapter, we pick out several topics related to the spin crossover that may be promising avenues for future applications of electronic devices and present the present progress and future perspective on the oxide materials different from organic molecular systems such as Fe complexes.

In summary, this book describes the historical arguments and recent experimental and theoretical developments of the spin-crossover phenomena in cobaltites beyond the classical treatment, as well as future perspectives on electronic applications. We hope that this book enables readers not only to clearly understand the background of the cobalt systems but also to enjoy the history and physics in the spin-crossover cobaltites.

Finally, we appreciate Prof. Shin-ichiro Iwai and Prof. Atsushi Fujimori for their encouragements at the early stage of this work and Dr. Akiyuki Tokuno and Ms. Taeko Sato for their efforts throughout the editorial procedures.

Tokyo, Japan — Yoichi Okimoto
Tokyo, Japan — Tomohiko Saitoh
Tokyo, Japan — Yoshihiko Kobayashi
Sendai, Japan — Sumio Ishihara

Contents

Chapter 1
Spin-Crossover Phenomena in Perovskite Cobaltites: Their History and Current Status of the Research

Yoshihiko Kobayashi, Keisuke Sato, and Kichizo Asai

Abstract The spin-crossover or spin-state transition and related phenomena in perovskite cobaltites have been attracting widespread interest among researchers for decades. One of the main issues has been the Co^{3+}-$(3d)^6$ spin-states in $LaCoO_3$. Despite extensive efforts of researchers, it has still been controversial as to what the magnetic excited state induced around 100 K is: intermediate spin (IS) with $S = 1$ or high spin (HS) state with $S = 2$. The IS state, which has a degree of freedom on e_g orbitals, is strongly supported by the experiments showing the Jahn–Teller distortion and/or orbital fluctuation. Several spectroscopic studies afford strong evidence for the HS state. However, HS is incompatible with the magnitude of observed magnetization and the ferromagnetic correlation among Co-spins. Therefore, neither spin-state model can comprehensively explain all the experimental facts. A heterogeneous spin phase including IS and HS as magnetic states has been proposed, which requires the presence of interaction among the magnetic Co spin-states. This idea, along with several recent experiments clarifying the collective nature of the spin-crossover phenomena, indicates that consideration beyond the $3d^6$ electronic state of single Co^{3+} is essential to understand the spin-crossover phenomena. The spin-crossover around 100 K is unique to $LaCoO_3$, whereas the spin-crossover accompanied by the insulator-to-metal transition (IMT) at around 500 K occurs commonly in $RECoO_3$ (RE: rare earth element). The fact also suggests a contribution of the nonlocal electronic state to the spin-crossover phenomena. In this review, we describe the history and current status of the research on the spin-crossover phenomena in perovskite cobaltites. We also review the anomalous magnetic and electronic properties such as the spin polaron and IMT at around 500 K, which are resulting from the electronic state responsible for the spin-crossover phenomena in $RECoO_3$.

Y. Kobayashi (✉)
Department of Physics, Tokyo Medical University, Shinjuku, Tokyo 160-8402, Japan
e-mail: koba@tokyo-med.ac.jp

K. Sato
Department of Industrial Engineering, National Institute of Technology, Ibaraki College, 866 Nakane, Hitachinaka, Ibaraki 312-8508, Japan

K. Asai
Department of Engineering Science, Graduate School of Informatics and Engineering, The University of Electro-Communications, Chofu, Tokyo 182-8585, Japan

Y. Okimoto et al. (eds.), *Spin-Crossover Cobaltite*, Springer Series in Materials Science 305, https://doi.org/10.1007/978-981-15-7929-5_1

Keywords Perovskite cobaltites · Spin-crossover or spin-state transition · Spin-states · Spin polaron · Insulator-to-metal transition

1.1 Introduction

The unusual physical properties on the dielectrics, magnetism and electrical transport, such as the multiferroics, superconductivity, metal–insulator transition, and colossal magnetoresistance have been extensively studied both theoretically and experimentally in transition metal oxides with the perovskite-type and related crystal structures [1–4]. Among them, trivalent cobaltites $RECoO_3$ (RE: rare earth element) have been attracting widespread interest of researchers on both basic science and applications mainly owing to their spin-crossover phenomena. In particular $LaCoO_3$, which is the typical perovskite cobaltite showing anomalous spin-crossover phenomena, has been widely studied since 1950s [5]. The material has been attracting interest because of its diverse properties accompanied by the spin-crossover phenomena, which are attributed to the combination among the multiple degrees of freedom on the spin, orbital, and lattice due to the $3d^6$ electronic configuration of Co^{3+}. In spite of the long history and the most extensive studies, unsolved riddles and controversy on the electronic-spin states responsible for the spin-crossover phenomena in $LaCoO_3$ have been remaining. In this review, we would feature the trivalent cobaltites with perovskite structure $RECoO_3$ to give an overview of its history and current status of the researches in the materials, and to explain what the origins of the difficulties and controversy are in clarifying its spin-crossover phenomena.

Before demonstrating the investigations of the physical properties and its electronic-spin state of $3d$ orbitals responsible for the spin-crossover phenomena in $RECoO_3$, we describe the characteristics of the perovskite structure and the electronic states of $3d$ orbitals based on the ligand-field theory, which has explained the electronic-spin states successfully in transition metal oxides and complexes.

The perovskite-type structure is common for many substances with a chemical formula ABO_3 [1]. The A-site accommodates cations such as RE and alkaline earths. Transition metal elements (M) often occupy the B-site. Figure 1.1 shows the crystal structure of perovskite for (a) cubic, (b) rhombohedral, and (c) orthorhombic unit cells. For cubic perovskite structure (space group: $pm\bar{3}m$), A ion is located at the cube corner position (0, 0, 0), B ion is located at the body center position (1/2, 1/2, 1/2), and O^{2-} ions are located at the face center position (1/2, 1/2, 0) and the equivalent positions as shown in Fig. 1.1a. Six ligand oxygen ions are cubic-symmetrically coordinated around B ion, which forms a BO_6 octahedron. The BO_6 octahedra make a three-dimensional array by sharing corner O^{2-} ions. The ideal cubic perovskite requires that the ionic radii of A, B, and O (r_A, r_B and r_O, respectively) satisfy the relation; $t = (r_A + r_O)/\sqrt{2}(r_B + r_O) = 1$, where t is known as the tolerance factor. In actual perovskite crystals, t often deviates from 1. In the case for $t < 1$, a tilting of BO_6 occurs, leading to a distortion from the cubic symmetry. $RECoO_3$ has a rhombohedrally distorted perovskite structure ($R\bar{3}c$) only for RE = La and

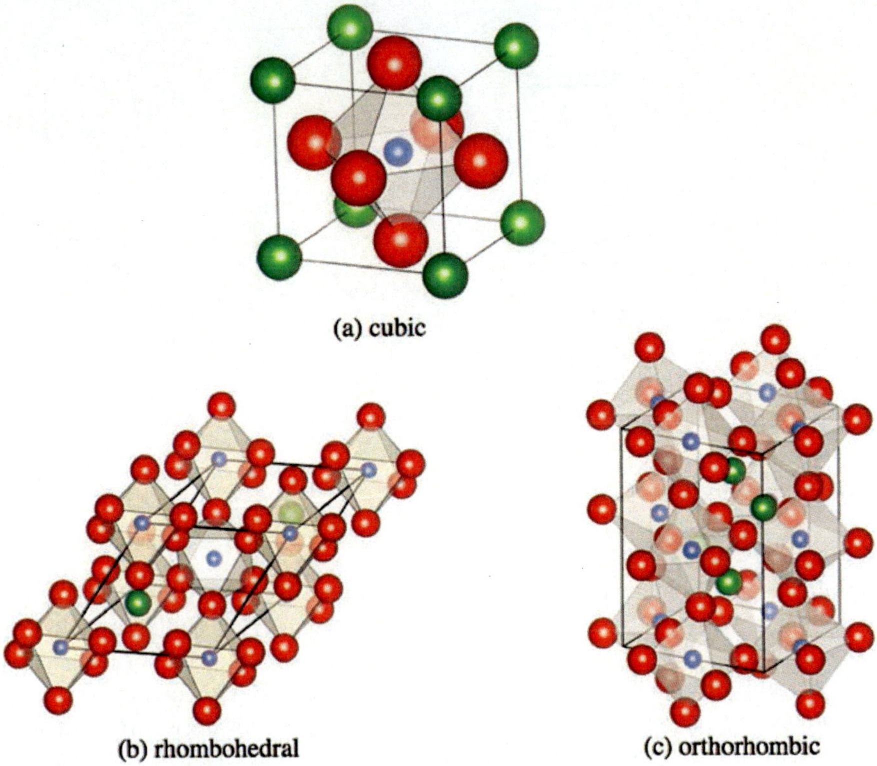

Fig. 1.1 Crystalline structure of perovskite ABO_3 for **a** cubic, **b** rhombohedral ($LaCoO_3$), and **c** orthorhombic unit cells ($RECoO_3$ with RE except La). The green, blue, and red balls denote A, B, and O^{2-} ions, respectively. The structural models are drawn using 3D visualization program for structural models: VESTA [6]

an orthorhombic structure (P_{nma}) for other RE at room temperature [see Fig. 1.1b, c respectively]. The unit cells are doubled and quadrupled for rhombohedral and orthorhombic structures, respectively. In case of $t > 1$, where large cations such as Sr^{2+} and Ba^{2+} occupy the A-site, the perovskite structure is often unstable and the layered structure with stacking of the face-shared BO_6 octahedra is realized [1]. The double perovskite structure $A_2B'B''O_6$ containing two kinds of B ions B' and B'' also exists, in which $B'O_6$ and $B''O_6$ octahedra are alternately arranged by sharing corner O^{2-} [1].

When the B-site accommodates the transition metal elements M with partially occupied d orbitals, the crystalline-electric field plays an important role in determining its electronic state. The ligand O^{2-} ions of MO_6 octahedron produce the crystalline-electric field with octahedral cubic symmetry at the transition metal site, leading to the splitting of the fivefold degenerated d electronic state of M ion into triplet t_{2g} orbitals: xy, yz and zx, and doublet e_g orbitals: $x^2 - y^2$ and $3z^2 - r^2$ [7], as shown in Fig. 1.2. The orbital angular momentum (**L**) responsible for the

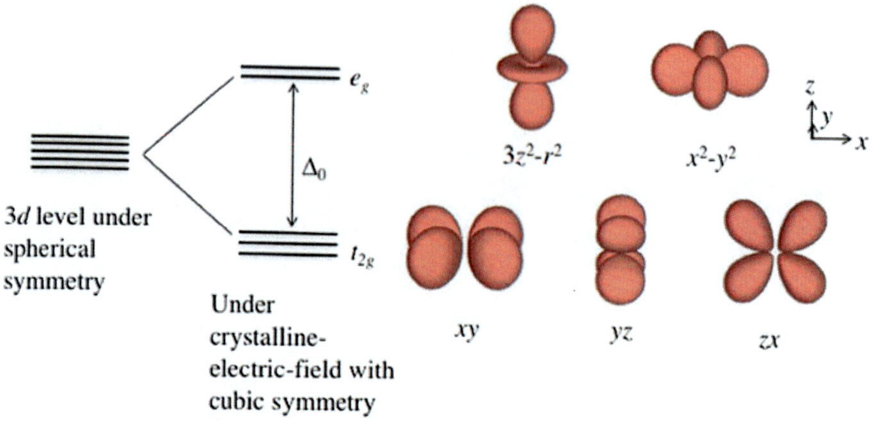

Fig. 1.2 Energy levels and orbitals of 3*d* electrons under the six-coordinated cubic-symmetrical crystalline-electric field

orbital magnetic moment vanishes for the e_g orbitals (quenching of the orbital angular momentum) and is much reduced for the t_{2g} orbitals. Thus, almost only the spin angular momentum remains. Note that the orbital angular momentum partially survives due to the spin–orbit interaction in actual materials, although the quenching of orbital angular momentum is a good approximation for *d* electron system (to be described later). The energy level for e_g is higher than that of t_{2g} orbital under the six-coordinated cubic-symmetrical crystalline-electric field, since the electronic wave functions of e_g orbitals extend to the ligand O^{2-} ions at the corner of MO_6 octahedron and are affected by a larger Coulomb repulsive interaction (see Fig. 1.2). Thus the *d* electrons in *M* ion under the six-coordinated cubic-symmetrical crystalline-electric field are preferentially accommodated in t_{2g} orbitals up to the *d* electron number $n = 6$. On the other hand, the intraatomic Coulomb repulsive interactions among the *d* electrons prefer the electrons to be placed separately in *d* orbitals so that the electrons keep away from each other. The intraatomic exchange interaction (Hund coupling), which originates from the Pauli exclusion principle and the intraatomic Coulomb repulsive interaction among the electrons, in general, leads to the Hund's rule; the configuration with the lowest energy is that with the maximum total spin *S*. Therefore, the realized electronic-spin configuration of the ground state for $4 \le n \le 7$ depends on the relative magnitude of the energy splitting due to the crystalline-electric field and the strength of the Hund coupling. When the energy difference among the electronic-spin states is small, multiple electron-spin configurations can be realized depending on the external environment such as the temperature, pressure, and the applied electric and magnetic fields. This phenomenon is called a **spin-crossover** or a **spin-state transition**.

In Co^{3+} ($3d^6$) compounds, the configuration $(t_{2g}\uparrow)^3(t_{2g}\downarrow)^3$, where all the electrons occupy t_{2g} orbitals, is expected to be the ground state for 3*d* electrons when the crystalline-field splitting predominates over the intraatomic exchange interac-

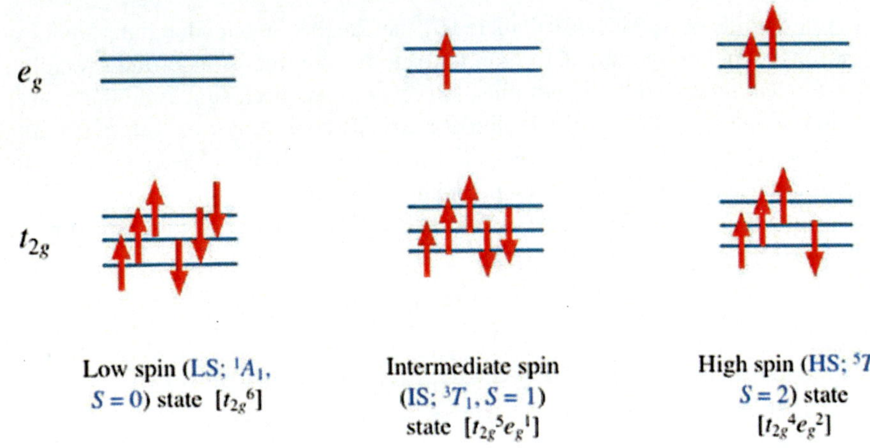

Fig. 1.3 Electronic-spin configurations for $3d^6$; Low spin (LS), Intermediate spin (IS), and High spin (HS) states

tion as shown in Fig. 1.3 (↑ and ↓ denote up and down spins, respectively). In this case, the **low spin (LS) state** ($^1A_1$) with the total spin $S = 0$ is realized whereas, $(t_{2g}\uparrow)^3(e_g\uparrow)^2(t_{2g}\downarrow)^1$ becomes the ground state when the intraatomic exchange interaction predominates, i.e., the **high spin (HS) state** ($^5T_2$) with the total spin $S = 2$ is realized. The **intermediate spin (IS) state** ($^3T_1$) with $(t_{2g}\uparrow)^3(e_g\uparrow)^1(t_{2g}\downarrow)^2$ and $S = 1$ is also possible as an electronic-spin configuration. The LS state with $S = 0$ is a singlet state with no magnetic moment owing to no degree of freedom on the orbital occupancy. The Co^{3+} with IS state, in which e_g orbitals accommodate one electron, is a Jahn–Teller (JT) ion. Both the HS and IS states have a degree of freedom on t_{2g} orbital, which allows the orbital angular momentum to remain in both states. The ionic radii expand by e_g orbital occupation since their electronic wave functions extend to the ligand O^{2-} ions, leading to $r_{LS} < r_{IS} < r_{HS}$ [8]. Tanabe and Sugano have conducted a systematic calculation of the energy level of $3d^n$ electronic-spin configurations under six-coordinated cubic-symmetrical crystalline-electric field as functions of the crystalline-electric field splitting relative to the intraatomic exchange interaction among the electrons [7, 9]. Their accomplishment based on the ligand-field theory is renowned as "Tanabe–Sugano diagrams", i.e., the energies of the electronic configurations are drawn as functions of the crystalline-field-splitting parameter [9]. The diagram for $n = 6$ predicts that the HS state is the ground state when the intraatomic exchange interaction is dominant ($\Delta_0/B < 20$), and the LS ground state is realized with increasing the crystalline-electric field. Here, Δ_0 and B denote the crystalline-field-splitting parameter and the Racah parameter, respectively [9].[1] The IS state cannot be the ground state according to the diagram (see Sect. 1.3.1).

[1]In [9], Tanabe and Sugano express the crystalline-field-splitting parameter as 10 Dq instead of Δ_0.

Note that the energy levels calculated by the Tanabe–Sugano diagrams are for the electronic-spin configurations of atomic (or ionic) *d* orbitals under the crystalline-electric field produced by ligand ions. However, it has been suggested that consideration taking into account the molecular orbitals consisting of *d* orbitals and the ligand *s* and *p* orbitals is necessary to discuss precisely the electronic-spin states responsible for the spin-crossover phenomena and the accompanying phenomena in real materials [10–12], which is relevant to the difficulty in understanding the spin-crossover phenomena in $LaCoO_3$.

1.2 Spin-Crossover and Accompanying Phenomena of $LaCoO_3$

Before focusing on the spin-crossover phenomena of $LaCoO_3$, we describe an overview of the physical properties of LaMO_3 for M = Fe and Ni, which are located on the left and right side neighbors of Co, respectively, in the periodic table. $LaFeO_3$ is a Mott insulator with an antiferromagnetic ordering of the localized magnetic moments below the Neel temperature T_N = 740 K, where $3d^5$ electrons on Fe^{3+} show the HS state with $(t_{2g}\uparrow)^3(e_g\uparrow)^2$ electronic-spin configuration (S = 5/2) [1, 2]. On the other hand, $LaNiO_3$ shows a metallic electrical conductivity with the Pauli paramagnetism. The Ni^{3+} with $3d^7$ has no localized magnetic moment in spite of the $(t_{2g}\uparrow)^3(e_g\uparrow)^1(t_{2g}\downarrow)^3$ electronic configuration (LS state; S = 1/2) or $(t_{2g}\uparrow)^3(e_g\uparrow)^2(t_{2g}\downarrow)^2$ electronic configuration (HS state; S = 3/2) according to the Tanabe–Sugano diagram, suggesting that the itinerant *d* electron view seems to be more suitable for describing the electronic-spin state of $LaNiO_3$ rather than the localized 3*d* electrons [1, 2]. From these circumstances, it is not surprising that $LaCoO_3$ exhibits various physical properties arising from complex electronic-spin states, which cannot be explained based on simple electronic-spin configurations predicted by the ligand-field theory. Next, we introduce the research outcomes of the magnetic and electric properties relevant to the spin-crossover phenomena of $LaCoO_3$ along its history.

Temperature dependence of magnetization and electrical resistivity of $LaCoO_3$ for a wide temperature range has been reported by Heikes et al.[13]. The magnetic susceptibility (χ) of $LaCoO_3$ increases, and decreases with increasing temperature after exhibiting a broad peak at approximately 100 K as shown in Fig. 1.4a, which has been considered to be an anomalous behavior for paramagnetic materials [14, 15]. (Note that the Curie–Weiss-like $\chi(T)$ at the lowest temperatures shown in Fig. 1.4a is due to magnetic impurities.) An unusual plateau is seen at approximately 500 K [see Fig. 1.4a]. The neutron scattering experiments have verified that the peak of $\chi(T)$ at approximately 100 K is not due to an antiferromagnetic ordering [16]. The electrical resistivity (ρ) shows a thermal-activation-type temperature dependence below approximately 300 K, exhibits a sharp drop of 2 or 3 orders in magnitude between 400 and 600 K suggesting an insulator-to-metal transition (IMT), and increases with

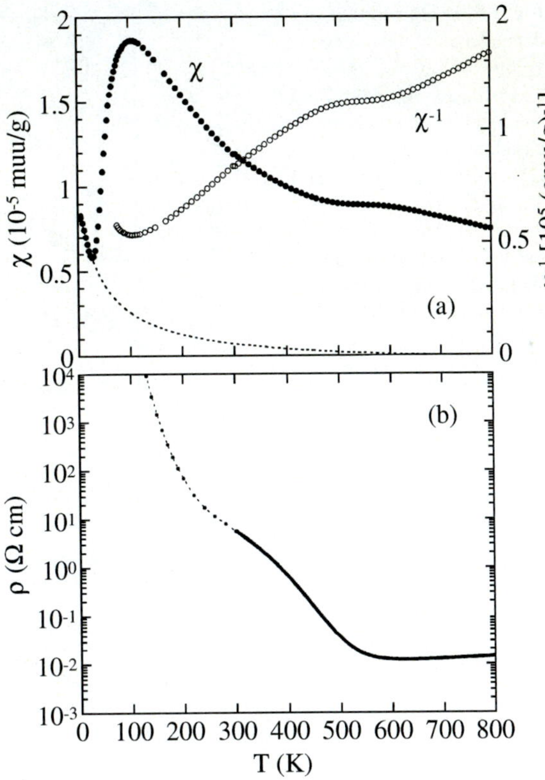

Fig. 1.4 Temperature dependence of **a** the magnetic susceptibility (χ) and reciprocal susceptibility (χ^{-1}), and **b** electrical resistivity (ρ) of $LaCoO_3$. The figures are redrawn using the data in [15]. The dotted line in Fig. 1.4a denotes the magnetic impurity component

increasing temperature above approximately 600 K [14, 15] as shown in Fig. 1.4b. That is, the magnetism of $LaCoO_3$ exhibits anomalous characteristics at approximately 100 and 500 K, and the latter accompanies a huge change of the electrical transport properties. The anomalies of magnetism at approximately 100 and 500 K are also accompanied by anomalous thermal lattice volume expansion as shown in Fig. 1.5 [14].

The neutron scattering experiments for wide ranges in both the reciprocal lattice space and temperature scale have been conducted by Asai et al. in order to obtain the information on the magnetic moment of Co and on its correlation [17]. Figure 1.6 shows the neutron magnetic scattering intensity at 295 K in the reciprocal lattice space of the pseudo-cubic unit cell of $LaCoO_3$. The neutron magnetic scattering intensity is spread in the whole reciprocal lattice space, suggesting the paramagnetic character of the Co magnetic moments. The neutron magnetic scattering intensity exhibits a broad maximum at around the scattering wave vector (100), which implies a weak ferromagnetic correlation among the magnetic moments with an extremely short correlation length. Figure 1.7 shows the temperature dependence of the paramagnetic neutron scattering intensity. The temperature dependence of the paramagnetic neutron scattering intensity mimics $\chi(T)$, i.e., the intensity, which is almost zero at

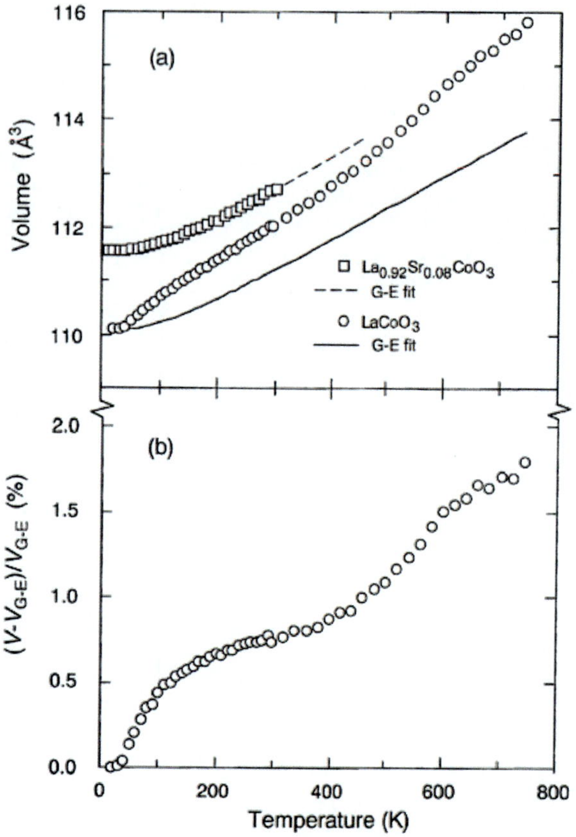

Fig. 1.5 **a** Temperature dependence of the lattice volume of $LaCoO_3$ and $La_{0.92}Sr_{0.08}CoO_3$ [14]. The solid and broken lines are based on the Grüneisen–Einstein model. **b** The anomalous part of the thermal expansion of $LaCoO_3$. (Reprinted from [14], Copyright (1998) The Physical Society of Japan.)

the lowest temperatures increases with increasing temperature and decreases after exhibiting a broad maximum near 100 K, although the magnetic anomaly at approximately 500 K cannot be seen within the experimental accuracy. This result definitely provides the evidence that the Co^{3+}, which has no magnetic moment at the lowest temperatures, has an increasing magnetic moment with increasing temperature.

The picture on the magnetism of the Co^{3+} in $LaCoO_3$ derived from the magnetization, thermal expansion, and neutron scattering experiments is as follows: Co^{3+} has a nonmagnetic LS ground state, and its averaged magnetic moment increases with increasing temperature most likely owing to the spin-crossover to some magnetic excited state above approximately 100 K, and another spin-crossover phenomenon occurs possibly above approximately 500 K. Although there have been large amounts of investigations claiming the magnetic–electric anomalies at approximately 100 K to be closely related to the spin-crossover phenomena, the common understanding on the magnetism of $LaCoO_3$ is nothing other than the issue described above. Hereafter, we call the change of the spin-state of Co-3d^6 in $LaCoO_3$ at approximately 100 K **"100 K-transition"**.

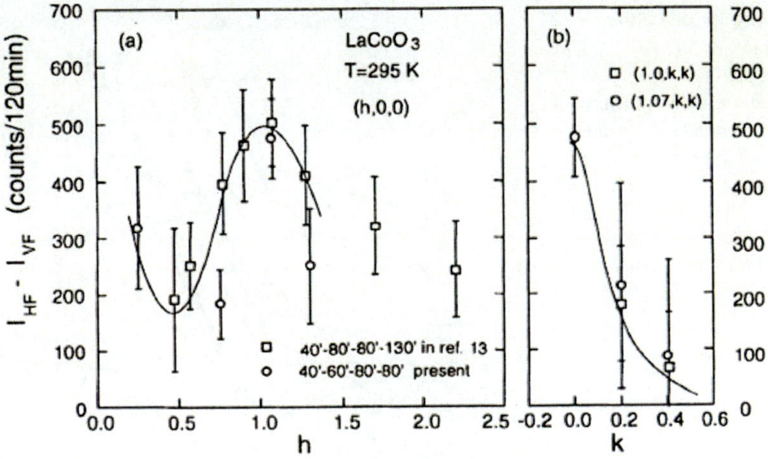

Fig. 1.6 Scattering vector dependence of the paramagnetic neutron scattering intensity at 295 K with $E_f = 41$ and $\Delta E = 0$ meV; **a** along $(h, 0, 0)$, and **b** along $(1.07, k, k)$ and $(1.0, k, k)$ for $LaCoO_3$ [17], where the indices denote those of the pseudo-cubic cell. (Reprinted from [17], with permission from Copyright (1994) by the American Physical Society.)

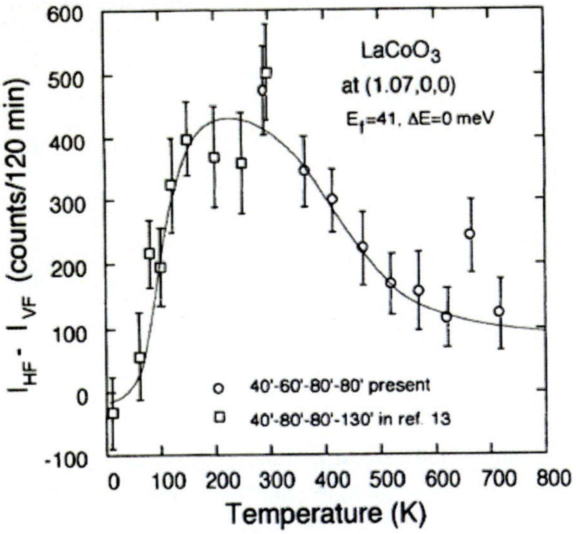

Fig. 1.7 Temperature dependence of the paramagnetic neutron scattering intensity for $LaCoO_3$ at (1.07, 0, 0) with $E_f = 41$ and $\Delta E = 0$ meV [17]. (Reprinted from [17], with permission from Copyright (1994) by the American Physical Society.)

In the next section, we would like to describe the long-time extensive efforts by many researchers to clarify the electronic-spin state of the magnetic excited state of $LaCoO_3$ above 100 K, and the conflicting arguments about its spin-states.

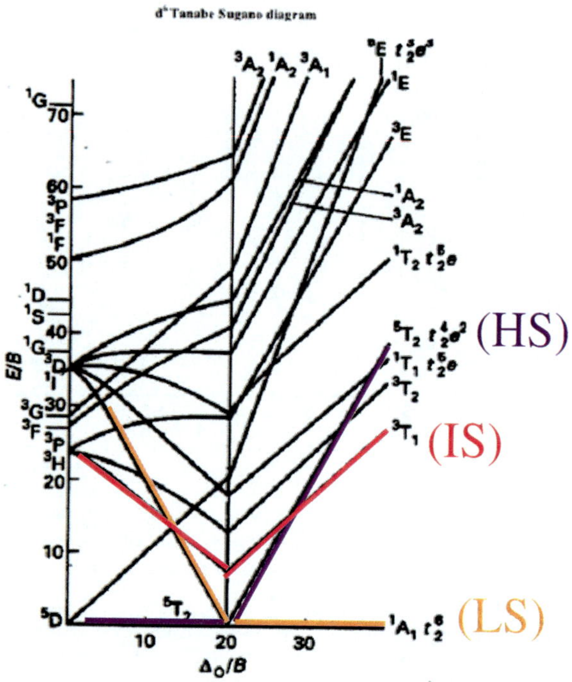

Fig. 1.8 Tanabe–Sugano diagram for $3d^6$ [9]. Δ_0 and B denote the crystalline-field-splitting parameter and the Racah parameter, respectively. The indications of low spin (LS), intermediate spin (IS), and high spin (HS) states were added for convenience to read.

1.3 The Controversy on the Spin-State of Co-3*d* Above 100 K-transition in $LaCoO_3$

In the early days of the research from 1960s to 1990s, the magnetic anomaly of $LaCoO_3$ at approximately 100 K was interpreted as due to the thermal excitation from LS ground state [$^1A_1$; $(t_{2g}\uparrow)^3(t_{2g}\downarrow)^3$] to HS excited state [$^5T_2$; $(t_{2g}\uparrow)^3(e_g\uparrow)^2(t_{2g}\downarrow)^1$]. This view was suggested by theoretical investigation on *d* electronic states based on the ligand-field theory [7, 9]; LS can be the ground state with HS state as the first excited state, when the crystalline-field splitting and the intraatomic exchange interaction compete with each other (Δ_0/B is slightly larger than 20), according to the Tanabe–Sugano diagram as shown in Fig. 1.8. This LS-to-HS transition model had, however, serious inconsistencies with the experimental results: (i) the magnitude of the magnetization expected from the HS state of Co^{3+} is much larger than the measured one [14, 15, 18] and (ii) the antiferromagnetic superexchange interaction between HS Co^{3+} expected from the Goodenough–Kanamori rule [19] contradicts with the ferromagnetic correlation revealed by the paramagnetic neutron scattering experiments in [17].

1.3.1 Experiments Supporting IS

In mid-1990s, several theoretical investigations suggesting the IS state [$^3T_1$; $(t_{2g}\uparrow)^3(e_g\uparrow)^1(t_{2g}\downarrow)^2$] as the excited state of the 100 K-transition were proposed. Potze et al. pointed out that the hybridization between Co-3d and O-2p orbitals stabilizes the IS state as the ground state for Co^{4+} with $(3d)^5$, and discussed the ferromagnetism of $SrCoO_3$ and the spin-glass phase of Sr-doped $LaCoO_3$ [10]. Korotin et al. calculated the electronic structure of $LaCoO_3$ using the LDA+U band structure calculations, and claimed that the IS state is the first excited state with much lower energy than that of the HS state [11]. They also discussed that the IS state of Co^{3+}, which is a JT ion with a degree of freedom on e_g orbital, can develop an orbital ordering below 500 K [11]. Mizokawa and Fujimori proposed that the IS state may play an important role in the 100 K-transition in $LaCoO_3$ based on a spin- and orbital-unrestricted Hartree–Fock band calculation [12]. Note that the hybridization between Co-3d and O-2p orbitals was the key mechanism for stabilizing the IS state in these calculations.

Following these theoretical studies, numerous experiments supporting LS-to-IS excitation as the 100 K-transition of $LaCoO_3$ have been successively reported. Saitoh et al. analyzed their Co-2p core-level photoemission, valence-band photoemission, and O-1s X-ray absorption spectroscopy (XAS) spectra, along with the temperature dependence of χ [18]. They clarified that the spectra can be interpreted as due to a gradual LS-to-IS transition, and also succeeded to reproduce measured $\chi(T)$ (see Chap. 2 by Saitoh in detail). Asai et al. demonstrated that the temperature dependence of the magnetization and the anomalous thermal lattice volume expansion can be explained using the LS, IS and HS states with the energy difference of ~200 K and ~1000 K, respectively, and proposed the two-stage spin-state transition model with the 100 K-transition from LS to IS and the second transition from IS to a mixed state of IS and HS (see Figs. 1.9, 1.10) [14]. The energy difference between the LS and IS ($\Delta = \sim$200 K) was also confirmed by the temperature dependence of the nuclear magnetic resonance (NMR) relaxation rates [20] and the elastic constants [21] (Fig. 1.10).

Yamaguchi et al. observed the splitting and intensity variation of phonon mode through the 100 K-transition using the infrared spectroscopy, and explained them as due to the local lattice distortion resulting from the JT effect of the thermally excited IS state of Co^{3+} [22]. Maris et al. conducted a high-resolution diffraction experiment using synchrotron radiation and obtained the structural transition with increasing temperature from rhombohedral $R\bar{3}c$ to monoclinic $I2/a$ through the 100 K-transition. They ascribed the symmetry change to Q_2-type JT distortion and the e_g-orbital ordering (see Fig. 1.11) owing to the LS-to-IS transition [23]. Ishikawa et al. observed a remarkable softening of the optical phonon modes coupled with the cooperative JT distortion using the Raman scattering, and interpreted the results as due to the e_g-orbital ordering [24]. The inelastic neutron scattering experiments also revealed the softening of the optical phonon modes with energies ~15 and 22 meV for a wide range of the reciprocal lattice space in contrast to no appreciable change in the acoustic phonons [25]. Thant et al. revealed an anomalous frequency dispersion of the

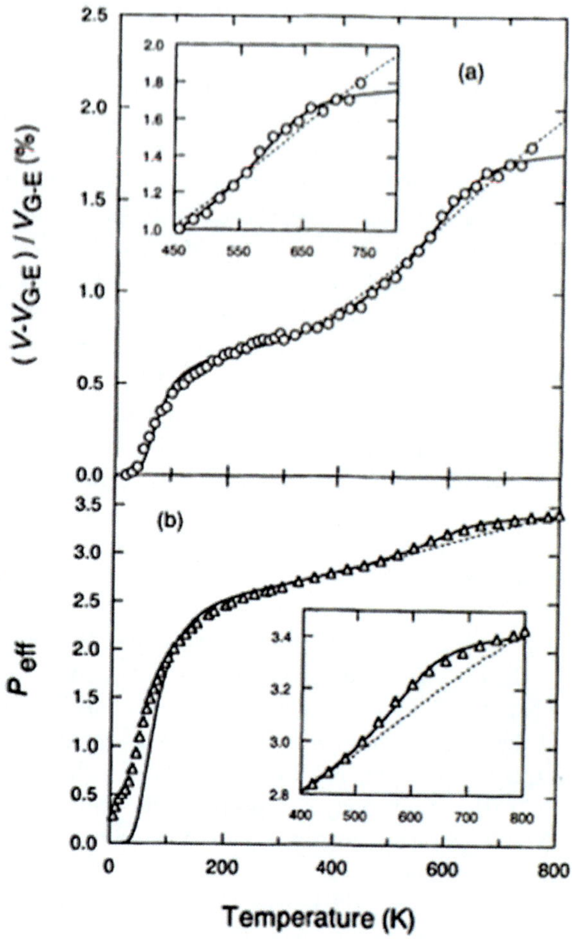

Fig. 1.9 Temperature dependence of **a** the anomalous part of the thermal expansion and **b** the effective magnetic moment (p_{eff}) of $LaCoO_3$, along with the fittings based on the two-stage spin-state transition model (solid lines) [14]. (Reprinted from [14], Copyright (1998) The Physical Society of Japan.)

sound velocity above 150 K for the ultrasonic frequency 10–100 MHz. They found that the lattice relaxation is thermal activation type, and derived the attempt frequency ~200 MHz and the activation energy ~120 K based on the Debye model [26]. In their report, the slow lattice relaxation was attributed to the e_g-orbital fluctuation of the IS state of Co^{3+}. Louca and Sarrao proposed a local static JT distortion induced by the magnetic excited state of Co ions based on the pair density function analysis of their neutron scattering data [27]. Phelan et al. observed a coexistence of the ferromagnetic and antiferromagnetic correlations by their inelastic neutron scattering experiments on $LaCoO_3$ and $La_{1-x}Sr_xCoO_3$ (x = 0.1–0.2). They attributed their results to the thermally excited IS state and dynamic orbital ordering [28].

The arguments supporting the IS as a magnetic state of Co^{3+} above 100 K are mainly based on the existence of JT distortion and/or e_g-orbital ordering in addition to the magnitude of magnetization and the ferromagnetic correlation.

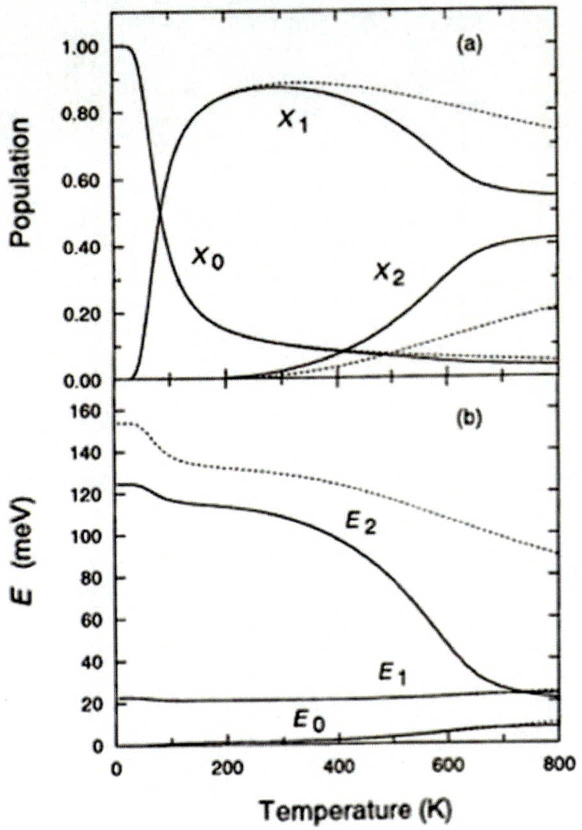

Fig. 1.10 Temperature dependence of **a** the population and **b** energies of LS, IS, and HS states according to the two-stage spin-state transition model [14]. (Reprinted from [14], Copyright (1998) The Physical Society of Japan.)

1.3.2 Experiments Supporting HS

Amid the progress in the research of the 100 K-transition based on the IS state of Co^{3+}, Radwanski and Ropka reported a theoretical calculation based on the ligand-field theory taking into account the spin–orbit interactions. According to their result, the excited state involving the 100 K-transition is a triplet with $S = 1$, which results from the HS state [32, 33]. Noguchi et al. proposed that the magnetic excited state induced at approximately 100 K is the triplet with the activation energy $\Delta = \sim 140$ K. They measured the electron spin resonance (ESR) spectra under a pulsed magnetic field up to 30 T with various frequency, temperature, and crystalline orientation. They analyzed their ESR spectra successfully using an effective spin-Hamiltonian with uniaxial anisotropy. The activation energy $\Delta = \sim 140$ K was estimated from the temperature dependence of the ESR intensity (see Fig. 1.12) [34].

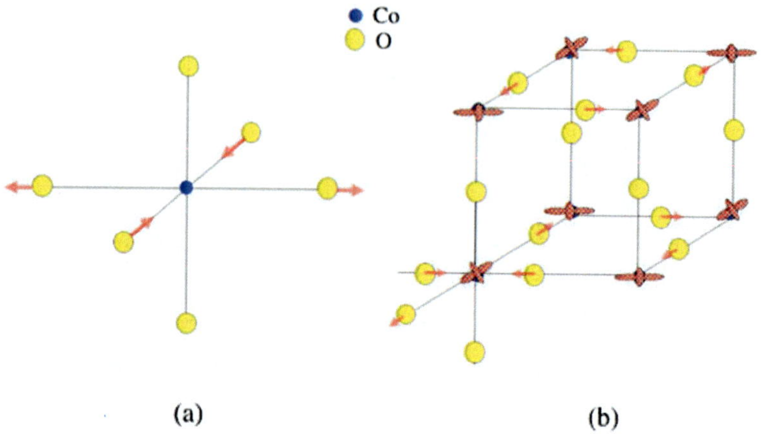

Fig. 1.11 **a** Q_2-type Jahn–Teller (JT) distortion [29] and **b** a possible ordering of e_g orbital. [30, 31]

Fig. 1.12 Energy diagram of the excited state of Co-3d in $LaCoO_3$ obtained by the ESR measurements [34]. The index [001] denotes that of the hexagonal unit cell. (Reprinted from [34], with permission from Copyright (2002) by the American Physical Society.)

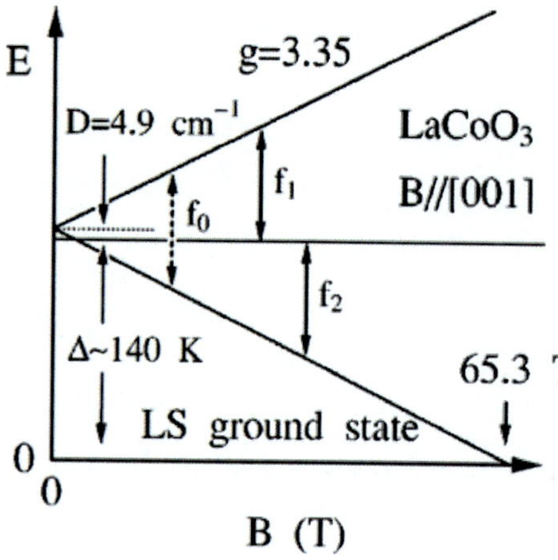

The g-values ($g_{\|} = 3.35$ and $g_{\perp} = 3.55$) and the zero-field-splitting $D = 4.90\,\mathrm{cm}^{-1}$ were obtained from the slope and the intercept of the magnetic field dependence of the resonance frequency, respectively. They attributed the strikingly large g-values to the triplet state derived from the HS state of 3d^6 in the six-coordinated cubic-symmetrical crystalline-electric field with the spin–orbit interaction. The obtained parameters in the ESR measurements such as the activation energy Δ, zero-field splitting D, and g-value were also supported by the inelastic neutron scattering experiments [28, 35] and the crystalline-electric-field calculation [36, 37]. The research by Noguchi et al. provided a strong experimental evidence supporting the HS as the excited state involved in the 100 K-transition. However, the magnitude of the magnetization expected from the activation energy $\Delta = \sim 140$ K and the large g-values exceeding 3.3

Fig. 1.13 Proposed arrangement of excited Co ions in the LS matrix of $LaCoO_3$ [39]. **a** HS Co ions are located as far apart as possible. **b** IS Co ions are collected to form domains. Open circles, solid circles, and solid squares indicate LS, HS, and IS state Co ions, respectively. (Reprinted from [39], with permission from Copyright (2005) by the American Physical Society.)

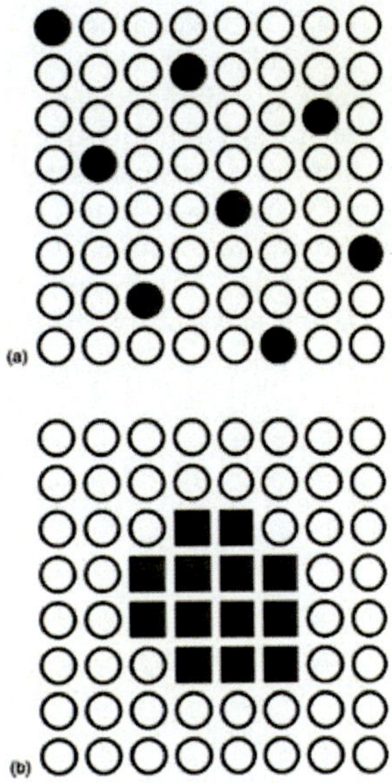

is about three times larger than the measured one. In order to solve the contradiction with the magnetization measurements, Kyômen et al. introduced a negative cooperative effect on the spin-state transition in $LaCoO_3$ [38, 39]. In their model, an energy difference between the LS ground state and the magnetic excited state increases by increasing the fraction of excited state owing to the repulsive interaction among the excited state Co ions [see Fig. 1.13a]. Using the model and parameters reported in [36], they reproduced the experimental data of the temperature dependence of the magnetic susceptibility and the heat capacity assuming the HS as the excited state as shown in Fig. 1.14. The excitation energy between the LS and HS is enhanced as the fraction of HS increases with temperature and reaches about 700 K at T = 300 K, whereas the excitation energy is ~200 K at low temperatures where most of Co^{3+} is in the LS state [see Fig. 1.14c]. They excluded the IS state as the excited state since a homogeneous mixture of the LS and IS is unstable and the IS-Co^{3+} tend to attract each other as shown in Fig. 1.13b, leading to a positive cooperative effect. Following these researches, a soft X-ray absorption spectroscopy and magnetic circular dichroism experiments [40], a comprehensive generalized gradient approximation (GGA+U) calculation [41], and a resonant inelastic X-ray scattering (RIXS) [42] were reported as studies supporting the model with the HS as the first excited state.

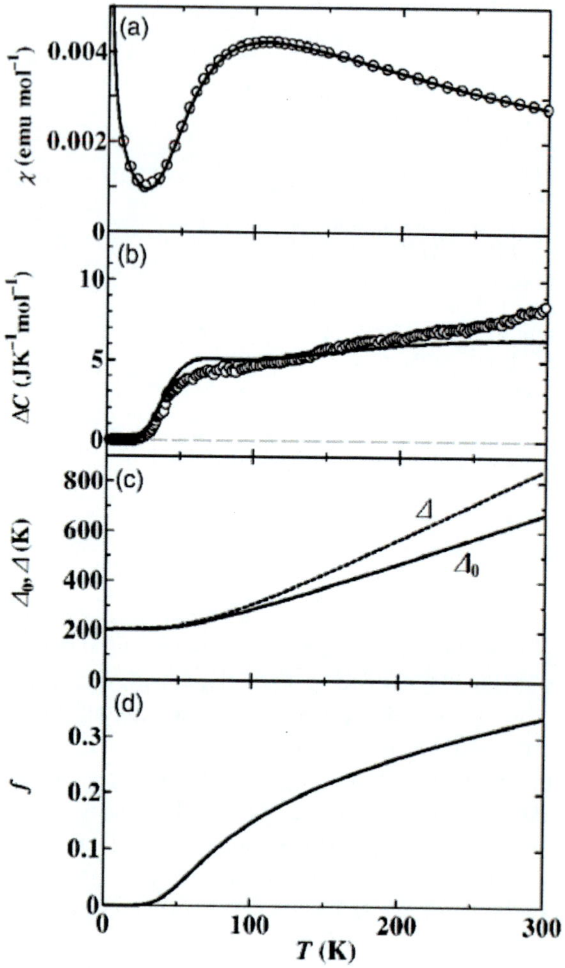

Fig. 1.14 **a** Magnetic susceptibility, **b** heat capacity, **c** net excitation energy, and **d** fraction of Co ions in the HS excited state [39]. (Reprinted from [39], with permission from Copyright (2005) by the American Physical Society.)

1.3.3 Coexistence of IS and HS, and Heterogeneous Spin Phases

Sato et al. proposed a coexistence of the IS and HS states based on the analysis of their high-field magnetization measurements using pulsed magnetic fields up to 67 T [43]. The field dependence of the magnetization exhibits a discontinuous increase with increasing magnetic field at around 60 T at 4.2 K as shown in Fig. 1.15, suggesting some magnetic transition. The transition field 60 T roughly agrees with the field of the level-crossing point shown in Fig. 1.12, where one of the Zeeman split levels of

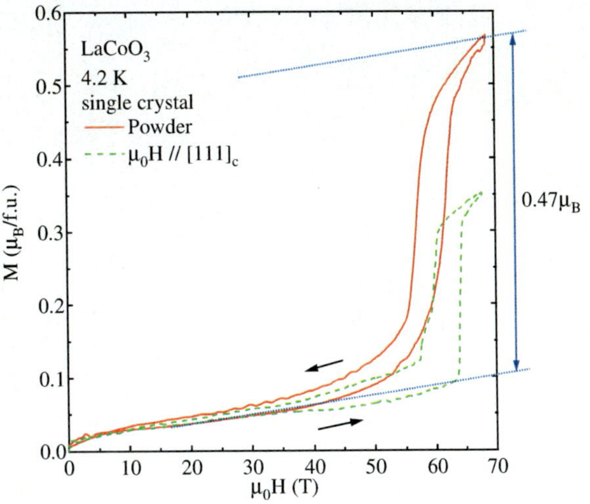

Fig. 1.15 High-field magnetization curves of single and powdered crystals of $LaCoO_3$ at 4.2 K [43]. (Reprinted from [43], Copyright (2009) The Physical Society of Japan.)

the triplet excited state intersects with the LS ground state. This fact implies that the field-induced spin-state transition occurs owing to the level crossing and supports the HS state as being induced above 60 T. The discontinuous increase of $M(H)$ accompanying a hysteresis suggests a cooperative nature of the field-induced spin-state transition resulting from some interactions between Co ions in the magnetic excited state. This fact is compatible with the above mentioned negative cooperative model assuming the repulsive interaction between the Co^{3+} in the HS excited state [38, 39]. However, the magnitude of the magnetization induced by the transition is 0.32 and 0.47 μ_B/Co for single crystal and powdered samples, respectively, which is much smaller than that expected for the triplet excited state with $S = 1$ and $g = 3.35$ [34, 36, 37, 39]. According to these results, Sato et al. proposed that a part of Co ions (less than 10 %) change the spin-state from the LS to HS at around 60 T with increasing magnetic field. They named the Co species contributing to the field-induced spin-state transition Co_{I} [43]. By assuming that the rest of Co ions (named Co_{II}) transform into the IS excited state ($S = 1$ and $g = 2$) with an activation energy ∼200 K, they successfully reproduced the temperature dependence of the magnetization measured at 7 T as shown in Fig. 1.16. (Note that they also assumed a magnetic impurity component named Co_{III}.) The parameters for calculating the magnetizations are shown in Table 1.1 [43].

Altarawneh et al. observed a two-stage field-induced magnetic transition at 63 and 73 T by the magnetization and magnetostriction measurements using pulsed magnetic fields up to 100 T [see Fig. 1.17a], and suggested the collective character of the spin-state transition. They assumed a field-induced magnetic excited state with $S = 1$ and $g = \sim 2$ [44]. In order to explain the collective behavior, they proposed a model including a repulsive interaction among the excited magnetic state of Co^{3+} between the nearest-neighbor and next-nearest-neighbor sites. In their model, the fraction of the magnetic Co^{3+} exhibits a step-wise increment from 0 to 1/4, 1/2, 3/4, and 1

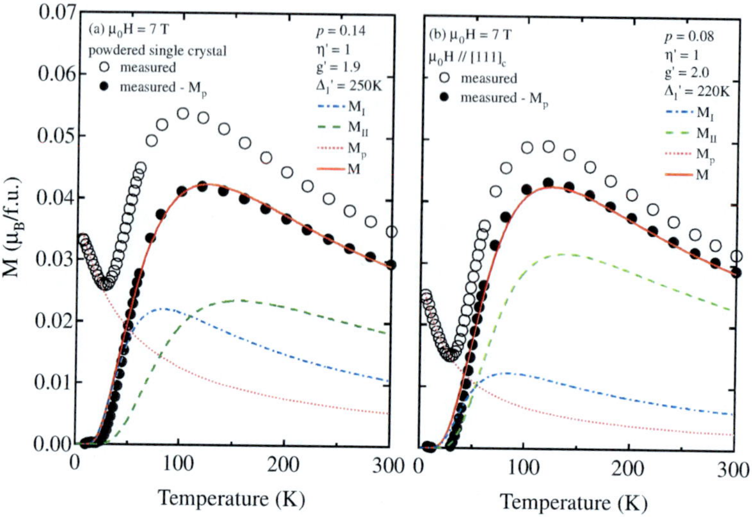

Fig. 1.16 Temperature dependence of the magnetizations at 7 T of **a** powdered single crystalline and **b** single crystalline $LaCoO_3$. The parameters for calculating the magnetizations are shown in Table 1.1 [43]. (Reprinted from [43], Copyright (2009) The Physical Society of Japan.)

Table 1.1 Parameters for the excited magnetic spin-state for the fitting of $M(T)$ in Fig. 1.16 along with those for other reports [43]. Δ, η, and g denote energy, orbital degeneracy, and g-value, respectively. Note that Ref. in the table represents the reference numbers in [43]. (Reprinted from [43], Copyright (2009) The Physical Society of Japan.)

Ref.		Δ (K)	η	g	Note
Present	Co_I	135	1	3.35	M
	Co_{II}	220–250	1	1.9–2.0	M
17		140	1	3.35	ESR
8		267	3	2	M
24		180	1	2.1	M
20		200–900	1	3.35	M, SH

M: magnetization, SH: specific heat

in succession with increasing field, with the critical fields H_{c1}, H_{c2}, H_{c3}, and H_{c4}, forming lattices of the LS and excited magnetic Co^{3+} as shown in Fig. 1.17b. Their model introduces the collective character of the field-induced spin-state transition, resulting in the heterogeneous spin phases including LS and magnetic Co^{3+} ions.

Ikeda et al. conducted high-field magnetization measurements on $LaCoO_3$ under a pulsed ultrahigh magnetic field of up to 133 T [45]. They discovered high-field phases and demonstrated a magnetic phase diagram for the field range 0–133 T and the temperature range 2–100 K, which is partly different from that by Altarawneh et al. [44]. The newly observed high-field phases are attributed to the field-induced

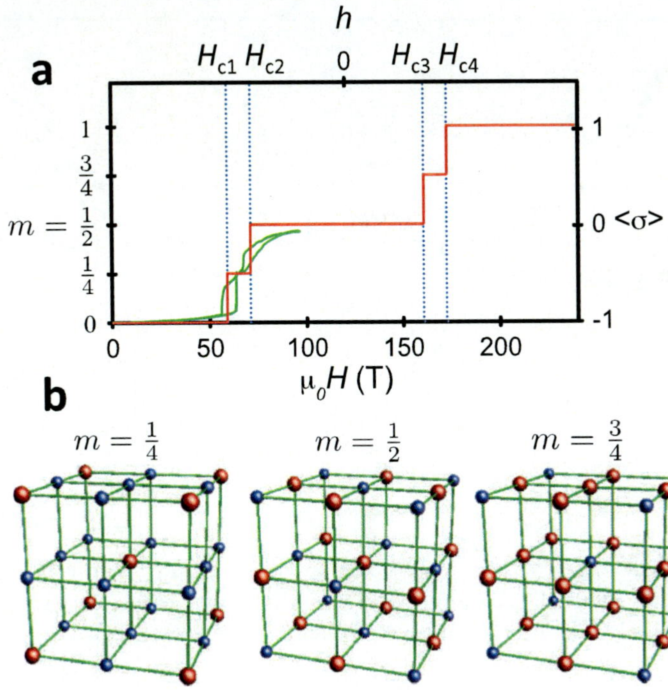

Fig. 1.17 **a** Field dependence of the magnetizations at 1.7 K (green line) along with the $M(H)$ predicted by the spin lattice model and **b** predicted spin-state crystalline structures for each magnetization plateau in panel (**a**) [44]. (Reprinted from [44], with permission from Copyright (2012) by the American Physical Society.)

excitonic insulator and LS/HS-ordered phases [46] (see the Chap. 3 by Ishihara for the details). Their results also support the presence of the heterogeneous spin phases.

The experimental study implying a coexistence of the HS and IS as magnetic species has been actually reported since the 1990s. The temperature dependence of the ^{59}Co-NMR Knight shift (K) in $LaCoO_3$ was reported by Itoh et al. [47–49]. The $K(T)$ of $LaCoO_3$ mimics its $\chi(T)$ as shown in Fig. 1.18a, in which only $LaCoO_3$ exhibits a peak at approximately 100 K corresponding to the 100 K-transition of Co^{3+}. In contrast to ordinary magnetic materials where $K(T)$ shows a linear behavior against $\chi(T)$, the K-χ plot of $LaCoO_3$ shows a deviation from a linear relation in the region of χ from 0 to ~4 [10^{-3} emu/mol] (the temperature range from 0 to ~90 K), and furthermore does not follow the trajectory with decreasing K and χ in the region of χ from ~4 to ~2 (the temperature range from ~90 to ~420 K), leading to drawing a loop as shown in Fig. 1.18b. At approximately 660 K, the K-χ plot overlaps with the first trajectory. The fact that the K versus χ relation is different in the two temperature regions suggests multiple magnetic states contributing to the K and χ. We consider that this anomalous K–χ relation is another support for the HS–IS coexistence model in which the fraction of the magnetic species varies with

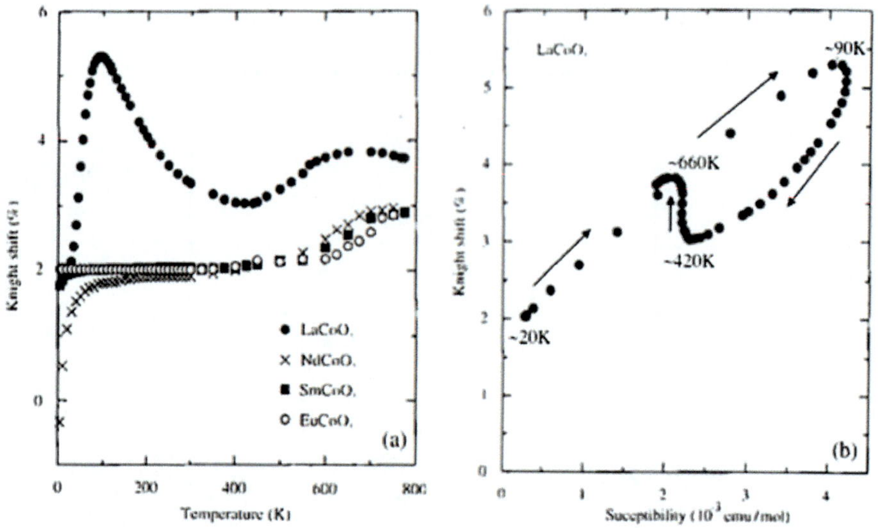

Fig. 1.18 **a** Temperature dependence of the ^{59}Co Knight shift in $RECoO_3$ (RE: La, Nd, Sm, and Eu) and **b** ^{59}Co Knight shift versus magnetic susceptibility plot with temperature as an implicit parameter in $LaCoO_3$ [48]. The temperature and allows in Fig. 1.18(b) are added by the authors of this article

changing temperature. (Note that the view described above is not compatible with that by Itoh et al.. They proposed the HS state as the magnetic excited state of $LaCoO_3$ based on the absence of the symmetry change due to the local JT distortion in [49].)

1.3.4 Summary of the Controversy

At the end of this section, we summarize the experimental results on the controversy about the magnetic state responsible for the 100 K-transition in $LaCoO_3$.

(i) The temperature dependence of the magnetic susceptibility along with the photoemission spectra is difficult to be explained by assuming the HS state $[(t_{2g}\uparrow)^3(e_g\uparrow)^2(t_{2g}\downarrow)^1]$ of Co^{3+}, and is reproduced assuming the IS state $[(t_{2g}\uparrow)^3(e_g\uparrow)^1(t_{2g}\downarrow)^2]$ as the magnetic excited state responsible for the 100 K-transition. These facts show that the IS state is dominant.
(ii) Several experiments exhibit local JT distortions on the Co-site and the e_g orbital ordering or fluctuation, supporting a partially occupied e_g state in Co^{3+}. The results cannot be explained assuming the HS state, which has no degree of freedom in e_g orbitals.
(iii) The magnetic excited state with the activation energy $\Delta = \sim 140\,K$ and $g = 3.35$, which was observed by the ESR and confirmed by the inelastic neutron scattering experiments, can be explained reasonably as the triplet state with $S =$

1 derived from the HS state. The triplet state with $S = 1$ had been proposed before by the crystalline-electric-field calculations taking into account the spin–orbit interaction.

(iv) In order to reproduce the results of the magnetization measurements by the triplet state derived from the HS state, a negative cooperative effect must be taken into account, i.e., an energy difference between the LS ground state and the magnetic excited state increases by increasing the fraction of the excited state owing to the repulsive interaction between the excited state Co ions. The cooperative behavior is consistent with a discontinuous $M(H)$ with hysteresis in the high-field magnetization experiments.

(v) The field-induced spin-state transition occurs at around 60 T, which coincides with the level-crossing field assuming $g = \sim 3$ and the activation energy $\Delta = \sim 140$ K proposed by the ESR and inelastic neutron scattering experiments, whereas it is revealed that only less than 10 % of Co^{3+} is excited to the magnetic state. The results along with the temperature dependence of the magnetization were interpreted as the rest of Co^{3+} ions maintain LS state at low temperatures and transform into IS excited state with increasing temperature, leading to coexistence of IS and HS states.

The experimental results summarized in items (i) and (ii) support the IS as the excited state for the 100 K-transition whereas, the results in item (iii) strongly suggest the existence of the excited state with $\Delta = \sim 140$ K and $g > 3.3$, which is most likely the HS state. The strong repulsive interaction among the HS proposed in item (iv) is consistent with the results in item (v), in which all the Co^{3+} never excited to the HS state at the same time. Thus, the magnetic excited state at 100 K-transition of $LaCoO_3$ does not have to be exclusively either IS or HS. The coexistence of IS and HS proposed in item (v) seems to be a possible solution to express the magnetic state of $LaCoO_3$, which resolves the controversy in the experimental results.

The spin-crossover phenomena in ordinary transition metal compounds and complexes have been well explained by the ligand-field theory of single transition metal ions [7, 9]. Thus, most of the researches presupposed the spin-states, LS, IS, and HS of a non-interacting Co^{3+} ions with definite energy levels for the 100 K-transition of $LaCoO_3$. However, if there is a strong interaction among the spin-states of neighboring Co^{3+} that exceeds the level difference of a single ion as the item (iv) proposes, it is inadequate to set the electronic spin-state of localized Co^{3+} ions as a premise to consider the electronic states responsible for the spin-crossover phenomenon in $LaCoO_3$. In order to treat the spin-crossover of $LaCoO_3$ precisely, the correlation among the Co-3*d* electronic-orbital states should be taken into account. In fact, theoretical studies such as a model calculation using a Dynamical Mean-Field Theory (DMFT) have been reported [50].

Several models emphasizing the nonlocal and/or itinerant character of the Co-3*d* electronic state have also been proposed very recently. One example is an excitonic insulator, which will be described in Chap. 3. The first-principle band calculation based on the Hartree–Fock mean-field approximation revealed a phase diagram with multi-Co-site spin-states, in which the Co moment is not uniform and has different values for each site, as a function of the Coulomb interaction and the lattice expansion [51]. In addition, several experimental facts have been reported, which suggest that the covalent bond among Co-$3d$, O-$2p$, and the orbitals of other elements is important for physical properties in Co oxides, as described later.

Through the accumulation of the research for the decades, it seems to be clear that the 100 K-transition is a collective and/or cooperative phenomenon, rather than an issue of localized $3d$ electrons on the Co ions. Consideration taking into account the collective character of the electronic structure is indispensable in order to understand the spin-crossover phenomena of $LaCoO_3$.

1.4 Element Substitution Effect—Flexibility of the Spin-State and Spin Polaron

1.4.1 Hole-Doping Effect

The substitution of elements in $LaCoO_3$ remarkably changes its magnetic and electrical transport properties. The most investigated is the substitution of divalent ions such as Sr^{2+} for La^{3+}, which leads to the introduction of Co^{4+} ions [1, 52].

The detailed magnetic phase diagram was proposed by Itoh et al. as shown in Fig. 1.19 [53]. In their phase diagram, the 100 K-transition disappears and the Co ions become magnetic down to the lowest temperature for $La_{1-x}Sr_xCoO_3$ with $x \geq 0.02$. This fact exhibits that not only the introduced Co^{4+} but the remaining Co^{3+} in $LaCoO_3$ have magnetic moments. They proposed a spin-glass-like ground state for $0.02 \leq x \leq 0.18$. The neutron magnetic scattering experiments revealed a ferromagnetic short-range correlation with a correlation length less than 10 Å for $x = 0.08$ [17]. A cluster-glass-like ferromagnetic phase appears in the range of $0.18 \leq x \leq 0.5$ [53]. Phelan found that the ferromagnetic correlation length increasing with increasing Sr content leads to a long-range ferromagnetic ordering for $x > 0.2$ [28]. Although the electronic state of the Co^{3+} ($3d^6$) in Sr-doped $LaCoO_3$ has been reported to be the same as the magnetic state for non-doped $LaCoO_3$ above 100 K [54], it has not been established yet. $\chi(T)$ shows a plateau at approximately 500 K for $x = 0.1$, suggesting that the 500 K spin-crossover remains for $La_{1-x}Sr_xCoO_3$ with $x \leq 0.1$ [15].

The electronic transport properties of $La_{1-x}Sr_xCoO_3$ have also been investigated well. The magnitude of the electrical resistivity ρ for $La_{1-x}Sr_xCoO_3$ with $x \leq 0.2$ is several orders smaller than that for $LaCoO_3$ below 500 K, although $\rho(T)$ remains to show a thermal-activation-type temperature dependence [$\rho(T) \propto \exp(E_g/k_BT)$] [15, 55]. The magnitude of the activation energy of the electrical conduction E_g remarkably decreases with increasing Sr content. The signs of the thermoelectric power (TEP) and Hall coefficients have clarified that the dominant carriers of the electrical conduction are positive holes, as expected from the substitution of divalent Sr^{2+} for La^{3+}. The IMT at approximately 500 K has been suggested to remain according to the temperature dependence of ρ and TEP measurements for $La_{1-x}Sr_xCoO_3$ with

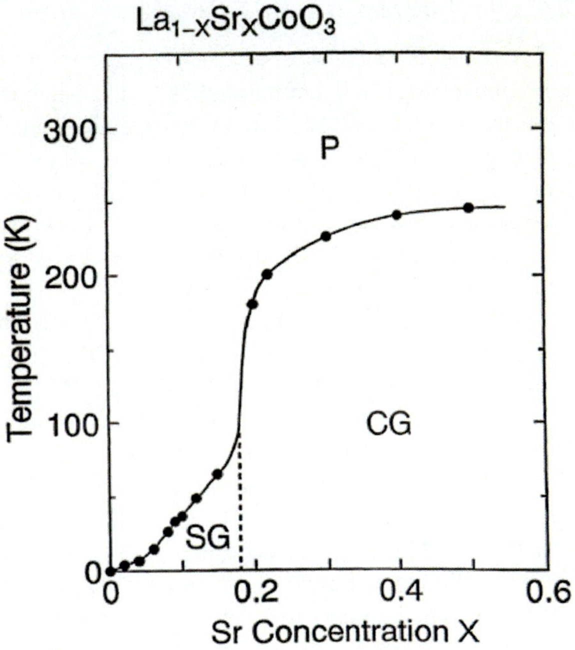

Fig. 1.19 Magnetic phase diagram of $La_{1-x}Sr_xCoO_3$ [53]. (Reprinted from [53], Copyright (1994) The Physical Society of Japan.)

$x \leq 0.2$ [15, 55]. For $0.2 \leq x \leq 0.5$, $La_{1-x}Sr_xCoO_3$ shows metallic ferromagnetic behavior below the ferromagnetic transition temperature T_c, which is attributed to a double exchange mechanism. Around T_c, a negative magnetoresistance was observed owing to a field-induced ferromagnetic alignment of the magnetic moments [56].

The substitution of Ca and Ba on $LaCoO_3$ causes similar change in the magnetic and the electrical transport properties [57–61]. The Ni-substitution effect to Co-site is similar to those of divalent ion substitution to La-site [15, 62, 63]. The TEP measurements clarified that positive holes are introduced in $LaCo_{1-x}Ni_xO_3$, possibly due to the introduction of Ni^{2+} in the Co^{3+}-site [15]. The heat capacity experiments for $La_{1-x}Sr_xCoO_3$ and $La_{1-x}Ca_xCoO_3$ exhibit that the electronic specific heat coefficient, null for undoped $LaCoO_3$, starts to increase sharply from $x = \sim 0.2$ for Sr and $x = \sim 0.25$ for Ca with increasing x, and saturates at higher concentrations for $x \geq 0.3$ [57]. This finding suggests the finite density of state at the Fermi level in the deeply substituted specimens. The similarity of the magnetic and the electrical transport properties in Ca-, Sr-, Ba-, and Ni-doped $LaCoO_3$ suggests that the hole-doping, not the ionic radius of the doped ions, dominates the physical properties of the substituted system.

A huge anisotropy in the magnetostriction [64, 65] and in the magnetization [66] was reported for $La_{1-x}Sr_xCoO_3$. The results were attributed to the IS state induced by hole-doping [64].

1.4.2 Electron-Doping Effect

In contrast to the hole-doping studies by divalent ion substitution, studies to introduce electrons in $LaCoO_3$ are rare possibly due to only a few suitable tetravalent ions. Gerthsen and Härdil reported the electron-doping effect by Th^{4+}-substitution with thorium content $x \leq 0.05$ on the electrical transport properties [55], in which the magnitude of ρ and TEP decrease with x. The electron-doping was verified by the negative TEP, although the decrease of ρ and TEP against thorium content is smaller in comparison with those with Sr content.

Fuchs et al. proposed a ferromagnetic order of the IS state Co spins induced by electron-doping in $La_{1-x}Ce_xCoO_3$ ($x \leq 0.4$) epitaxial thin films [67]. However, no one has successfully synthesized single-phase $La_{1-x}Ce_xCoO_3$ in bulk form, which is possibly due to the large difference in the ionic radius between La^{3+} and Ce^{4+} (1.16 and 0.97 Å for 8 coordination, respectively [8]). In order to deal with the problem on ionic radius mismatch, Kobayashi et al. chose $YCoO_3$ as a host material for Ce^{4+} substitution. The ionic radius of Ce^{4+} is much smaller than that of La^{3+} but close to that of Y^{3+} (1.02 Å for 8 coordination) [8]. They investigated the electron-doping effect on the magnetic and electronic transport properties of bulk $Y_{1-x}Ce_xCoO_3$ ($x \leq 0.1$) [68]. The introduced Ce ions were identified to be tetravalent by the hard X-ray photoemission and X-ray absorption spectroscopy experiments. The magnetization measurements suggest that the doped electrons induce HS Co^{2+} ($S = 3/2$) of the same amount as the Ce content and that the remaining Co^{3+} ions are in LS state. This behavior is quite different from that of the hole-doped $La_{1-x}Sr_xCoO_3$. The ferromagnetic ordering was observed below 7 K for $x = 0.1$. The feature of decrease in magnitude of ρ and negative TEP with x was consistent with those of thorium-doped $LaCoO_3$. More insulating behavior in comparison with the hole-doped system was explained as due to the cancelation of the doped electrons with the ligand holes. The smallness of the number of induced magnetic Co ions and the weak ferromagnetic interaction by the electron-doping, in comparison to the hole-doped system, were attributed to the more localized character of the doped carriers, which results in no magnetic cluster formation such as spin polaron and the weakness of the double exchange interaction [68].

Tomiyasu realized an electron-doping in $LaCoO_3$ by Te^{6+} substitution for Co^{3+} [69, 70]. They argued that the magnitude of Co moment induced by the electron-doping is almost the same as that of non-doped $LaCoO_3$ above 100 K since the Curie constant estimated from the Curie–Weiss fitting of $\chi(T)$ hardly depends on Te content. Based on their results, they suggested that HS Co^{3+} in $LaCoO_3$ is mainly replaced by HS Co^{2+} induced by the electron-doping. The measurements of the magnetization curves and the inelastic neutron scattering revealed that the magnetic correlation among the Co moments is very weak in the electron-doped $LaCoO_3$, in contrast to the hole-doped $LaCoO_3$. The decrease of ρ by the electron-doping in Te-doped $LaCoO_3$ is much smaller than that by the hole-doping, which is similar to that of above mentioned $Y_{1-x}Ce_xCoO_3$. These magnetic and the transport properties were explained based on the spin-state blockade mechanism [69, 70].

1.4.3 Lightly Doped $LaCoO_3$—Spin Polaron

Recently, a nanoscale phase separation in lightly hole-doped $LaCoO_3$ has been proposed to be important for the magnetic and transport properties of the substance.

In early days, Yamaguchi et al. proposed that the hole-doping with $x < 0.01$ in $LaCoO_3$ leads to form localized magnetic polarons with unusually large spin (S = 10–16) [71]. Podlesnyak et al. investigated the magnetic property of $La_{1-x}Sr_xCoO_3$ with $x = \sim 0.002$ using the inelastic neutron scattering, ESR and NMR measurements [72, 73]. They observed a magnetic scattering at 0.75 meV, which has large effective g-value ∼10–18. They proposed a formation of octahedrally shaped "spin-state polarons" comprising seven Co ions, in which one LS Co^{4+} ($S = 1/2$) and six IS Co^{3+} ($S = 1$) form a heptamer with the magnetic moment 13 μ_B, as shown in Fig. 1.20 [73]. Podlesnyak et al. also proposed that higher hole-doping with $x \geq\sim 0.05$ leads to a decay of spin-state polarons and a formation of larger scale magnetic clusters by ferromagnetic interactions [74].

Sato et al. investigated the lightly doping effect ($x \leq 0.05$) on the magnetism of $LaCoO_3$ by high-field magnetization experiments (see Fig. 1.21), and analyzed the data using the model to analyze the results of the field-induced spin-state transition of non-doped $LaCoO_3$ described in Sect. 1.3.3 [75, 76]. Based on the results (see Fig. 1.22), they proposed a spin-phase separation and a spin polaron formation in $La_{1-x}Sr_xCoO_3$, where about 30 nonmagnetic Co ions of two species (Co_I and Co_{II} described in Sect. 1.3.3) are converted to magnetic Co ions (Co_{III} described in Sect. 1.3.3) by introducing one Sr^{2+} at the lowest temperature as shown in Fig. 1.22c, leaving the remaining Co_I and Co_{II} in nonmagnetic ground state being only slightly affected by the Sr-substitution. The number ∼30 suggests that the Co^{3+} on the nearest and second nearest sites of Co^{4+} change to be magnetic Co_{III}. The length to the second nearest Co-sites is close to the ferromagnetic correlation length ∼8 Å for $La_{0.92}Sr_{0.08}CoO_3$ observed in the neutron magnetic scattering experiments [17]. The analyses of the field and temperature dependence of the magnetization revealed that the magnitude of the magnetic interaction within the clusters formed by Co_{III} is robust (larger than 10 T) while the magnetic interactions among the clusters responsible for the spin-glass formation are weak enough to be decoupled by the external field [75, 76]. By Rh^{3+}- and Ir^{3+}-substitution for Co^{3+}, nonmagnetic Co species were also found to convert to the magnetic Co_{III}, despite that the substitution of the trivalent ions formally introduce neither Co^{4+} nor carrier. The introduction of Rh^{3+} and Ir^{3+} induces about 1.3 and 11 Co_{III} per one substitution, respectively, as shown in Fig. 1.22a, b [75, 76]. This suggests that the efficacy to produce Co_{III} is much less than that of the hole-doping (∼30 Co_{III}/Sr^{2+}), as argued also in [77]. Ir^{3+} has about 10 times larger ability to convert the spin-state than Rh^{3+}, which is possibly due to the larger ionic radius of Ir^{3+} than that of Rh^{3+} (0.68 and 0.665 Å for Ir^{3+} and Rh^{3+}, respectively), or larger hybridization effect with Co-3d resulting from the more spatially extended Ir-5d state than Rh-4d. On the spin-state conversion mechanism by introducing trivalent 4d and 5d transition metal ions, several models taking into account the electronic structure have been proposed [78–80].

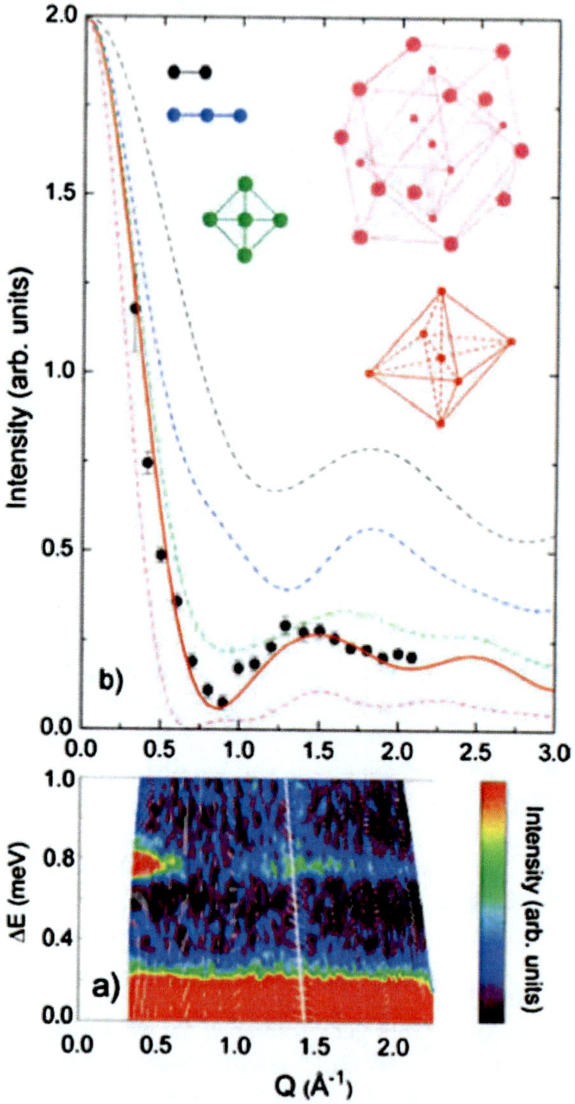

Fig. 1.20 **a** Inelastic neutron scattering spectrum of $La_{0.998}Sr_{0.002}CoO_3$ at 1.5 K. **b** Calculated Q-dependence of neutron cross section for different Co multimers (see insets), along with the measured Q-dependence of peak intensity at 0.75 meV [73]. The octahedrally shaped "spin-state polarons" consisting of seven Co ions reproduce the measured Q-dependence of peak intensity. (Reprinted from [73], with permission from Copyright (2008) by the American Physical Society.)

Tomiyasu et al. investigated Sc-substitution on the Co-site of $LaCoO_3$ [81]. Sc^{3+} has no d electrons and larger ionic radius (0.75 Å for 6 coordination) than that of HS Co^{3+} [8]. The lack of d orbitals breaks the magnetic interaction among Co^{3+} spins and reduces the hybridization between Co-3d and O-2p on average. The larger ionic radius of Sc^{3+} promotes emerging the HS or IS Co^{3+} with a larger ionic radius than the LS Co^{3+}. Thus, Sc-substitution on the Co-site of $LaCoO_3$ are expected to suppress the itinerant character of Co-3d electrons and at the same time to promote the localized magnetic Co^{3+}. The magnetization of Sc-doped $LaCoO_3$ shown in [81] exhibits that the LS ground state is suppressed and the Curie–Weiss-like $\chi(T)$ appears

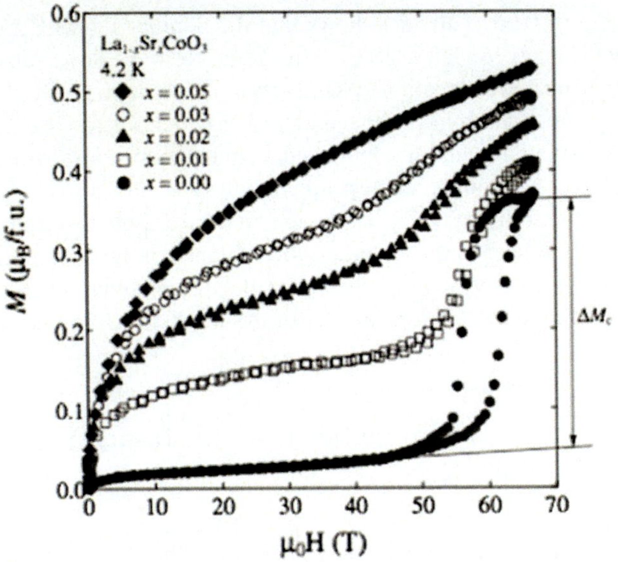

Fig. 1.21 **a** Field dependences of the magnetization of $La_{1-x}Sr_xCoO_3$ with $x = 0.00, 0.01, 0.02, 0.03, 0.04$, and 0.05 [75]. (Reprinted from [75], Copyright (2011) The Physical Society of Japan.)

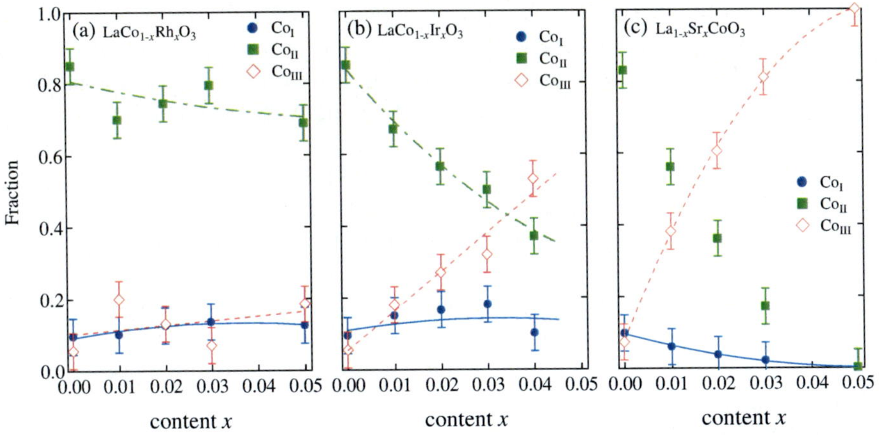

Fig. 1.22 x dependence of the fractions of Co_I, Co_{II}, and Co_{III} for **a** $LaCo_{1-x}Rh_xO_3$, **b** $LaCo_{1-x}Ir_xO_3$, and **c** $La_{1-x}Sr_xCoO_3$ [76]. (Reprinted from [76], Copyright (2014) The Physical Society of Japan.)

for Sc content of 0.05. This fact shows that the nonmagnetic LS Co^{3+} is partially replaced by the magnetic HS or IS Co^{3+} resulting from the Sc^{3+}-substitution, which partly mimics the above mentioned Rh^{3+}- and/or Ir^{3+}-doping effect.

The spin polaron and spin-phase separation, in which a part of Co^{3+} ions change to magnetic by a small amount of doping and form magnetic clusters in a nonmagnetic matrix of LS Co^{3+}, strongly suggest the flexibility of the spin-state of Co^{3+} ($3d^6$). The change of the spin phase of $LaCoO_3$ arising from a slight disturbance implies that the spin phase is determined by a quite subtle balance of the condition in the system. Such properties are hardly expected from the electronic-spin state based on the single Co ion, supporting a collective character of the 100 K-transition in $LaCoO_3$.

1.5 Spin-State Transition and Insulator-to-metal Transition at Around 500 K

As mentioned before, $LaCoO_3$ exhibits a decrease of the electrical resistivity by three orders in magnitude at around 500 K with increasing temperature [see Fig. 1.4b], which suggests an IMT. Tokura et al. reported that the optical conductivity spectra of $LaCoO_3$ shows a gap feature at 9 K and the onset of the conductivity (< 0.5 eV) remarkably shifts to lower energy with increasing temperature at around 500 K as shown in Fig. 1.23, i.e., a steep closing of the charge gap takes place in this temperature region. They pointed out that the feature of the optical conductivity spectra mimics those of the Mott transition induced by the filling-control in the strongly correlated electron systems, and attributed their results of $LaCoO_3$ to the thermally induced Mott transition [82]. The IMT has been observed in $\rho(T)$ of $RECoO_3$ with RE = La, Pr, Nd, Sm, Eu, Gd, Ho, Lu, and Y, where the transition temperature T_{IMT} depends on RE (for example, T_{IMT} =∼630 K for Gd), as shown in Fig. 1.24 [83–85].

For RE = La, Pr, Nd, Eu, Gd, Dy, Lu, and Y, an anomalous lattice volume expansion has been observed at around the temperature where the large change of $\rho(T)$ takes place [14, 83, 86–88]. These results imply that the IMT accompanied by anomalous lattice volume expansion at around 500–700 K is a common feature of $RECoO_3$. The anomalous part of the lattice volume expansion in the temperature region is about 4 % except for RE = La [83, 86–88]. For RE = La, the lattice volume expansion is only ∼ 1% at around 500 K [see Fig. 1.25] [14]. Note that $LaCoO_3$ exhibits an anomalous lattice volume expansion about 1% associated with the 100 K-transition as shown in Fig. 1.9a.

In $RECoO_3$, magnetic anomalies are also observed at around T_{IMT}. In $LaCoO_3$, a plateau at approximately 500 K in addition to the anomaly at approximately 100 K in the temperature dependence of magnetic susceptibility $\chi(T)$ has been reported as previously described (see Fig. 1.4) [14]. On the other hand, an increment of $\chi(T)$ has been reported for RE = Y and Lu only at the temperature region 500–700 K [83]. The effective magnetic moment (p_{eff}) for RE = La, Y, and Lu estimated from $\chi(T)$

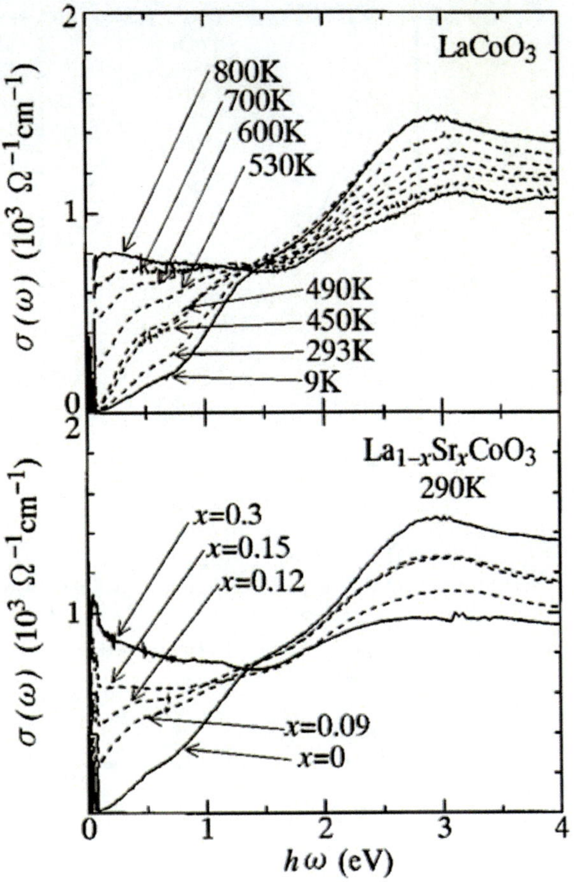

Fig. 1.23 Optical conductivity spectra for $LaCoO_3$, along with those of $La_{1-x}Sr_xCoO_3$ [82]. (Reprinted from [82], with permission from Copyright (1998) by the American Physical Society.)

assuming the Curie law increases steeply with increasing temperature at around T_{IMT} in a similar manner to the lattice volume as shown in Fig. 1.25 [14]. The temperature dependence of the ^{59}Co-NMR Knight shift K of $RECoO_3$ (RE = La, Nd, Sm, and Eu) also shows a step-like increment with increasing temperature at around T_{IMT} as shown in Fig. 1.18a [48]. These facts clarify that the magnetic moment of Co^{3+} increases with increasing temperature at around T_{IMT} in all $RECoO_3$, implying that a change of the spin-state of Co-3d accompanied by the IMT is a common feature of $RECoO_3$. Hereafter, we call the spin-state transition associated with the IMT in $RECoO_3$ **500 K-transition**, although T_{IMT} ranges from approximately 500 to 700 K depending on RE.

Kobayashi et al. investigated the crystalline structure and the magnetic properties of $La_{1-x}Pr_xCoO_3$ [89]. Temperature showing the "100 K-transition" of $LaCoO_3$ increases with increasing Pr content, and merges into that of the 500 K-transition for $PrCoO_3$ as shown in Fig. 1.26. The energy difference, Δ_1, between LS and IS states estimated from $p_{eff}(T)$ of Co^{3+}, assuming the two-stage spin-state transition model

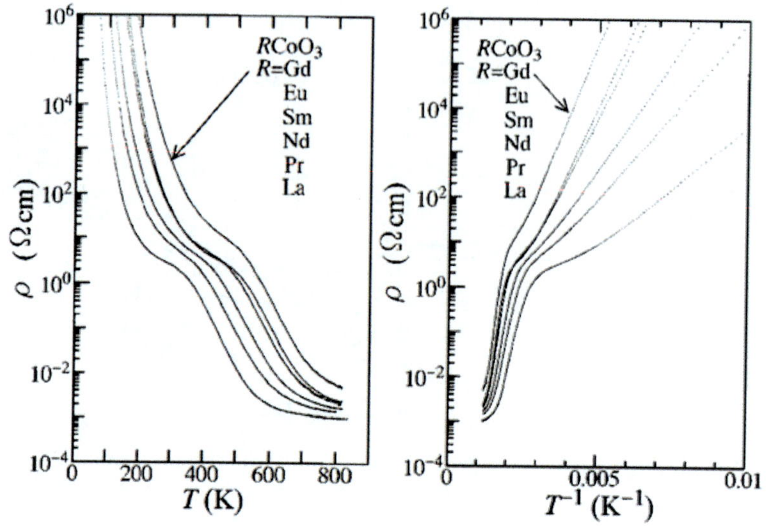

Fig. 1.24 $\rho(T)$ (left) and $\rho(T^{-1})$ (right) for $RECoO_3$ (RE = La, Pr, Nd, Sm, Eu, and Gd. [85]. (Reprinted from [85], with permission from Copyright (1996) by the American Physical Society.)

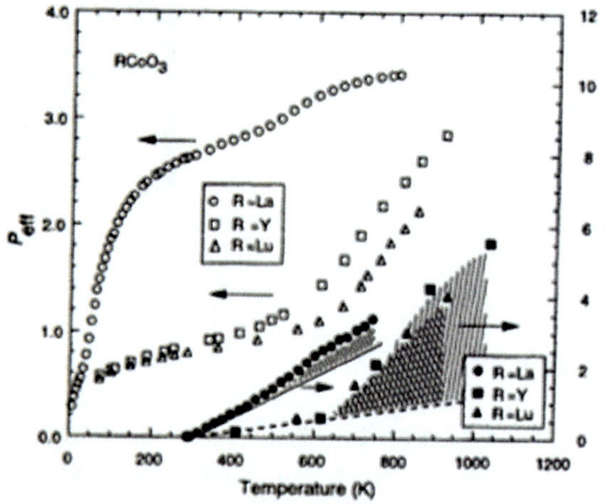

Fig. 1.25 Temperature dependence of p_{eff} and the lattice volume of $RECoO_3$ with RE = La, Y, and Lu [14]. (Reprinted from [14], Copyright (1998) The Physical Society of Japan.)

(LS–IS–HS transition model) and the Curie law, increases with increasing Pr content through the chemical pressure resulting from the smaller ionic radius of Pr^{3+} (1.13 Å for 8 coordination) in comparison with that of La^{3+} [8]. In contrast, the energy difference, Δ_2, between LS and HS states is almost independent of the Pr content (see Fig. 1.27).

Lightly hole-doped $LaCoO_3$ ($La_{1-x}Sr_xCoO_3$ and $LaCo_{1-x}Ni_xO_3$ with $x <\sim 0.1$) exhibits the 500 K-transition with IMT as shown in Figs. 1.28 and 1.29 [15, 55], although the substance shows no 100 K-transition and has magnetic ground state.

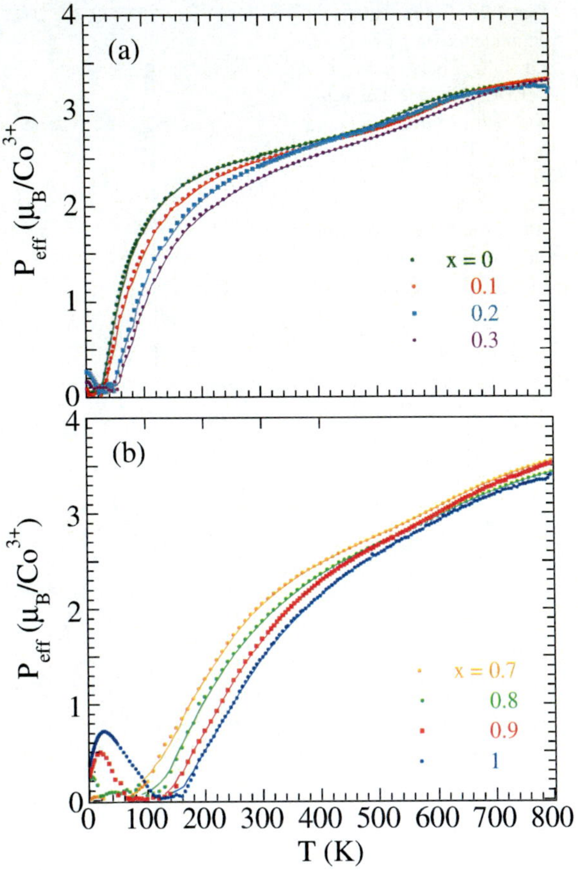

Fig. 1.26 Temperature dependence of p_{eff} of Co^{3+} in $La_{1-x}Pr_xCoO_3$ [89]. p_{eff} at low temperatures is a fictitious one caused by insufficient subtraction of the Pr moment. (Reprinted from [89], Copyright (2006) The Physical Society of Japan.)

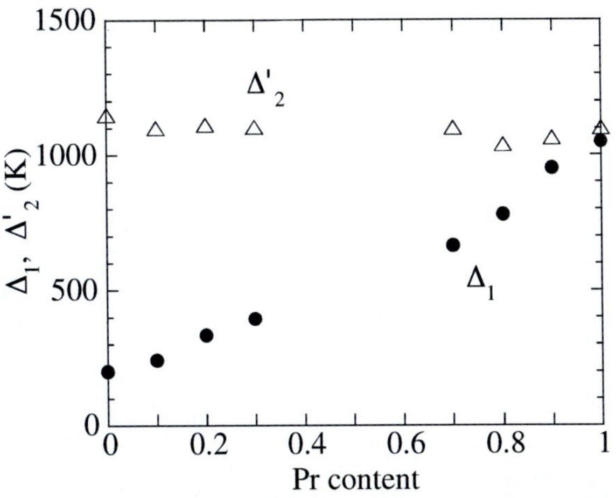

Fig. 1.27 Pr-content dependence of the energy levels between LS and IS, and LS and HS states (Δ_1 and Δ'_2, respectively) [89]. (Reprinted from [89], Copyright (2006) The Physical Society of Japan.)

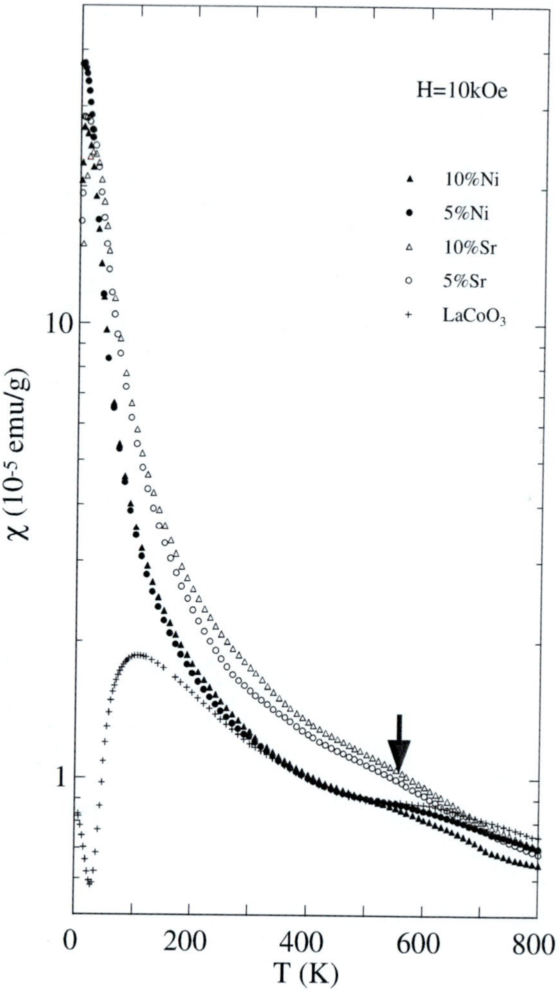

Fig. 1.28 Temperature dependence of χ of $LaCoO_3$, $La_{1-x}Sr_xCoO_3$, and $LaCo_{1-x}Ni_xO_3$. The figure was redrawn using the data in [15]. The arrow shows the magnetic anomalies responsible for the 500 K-transition. Note that the vertical axis is drawn on a logarithmic scale

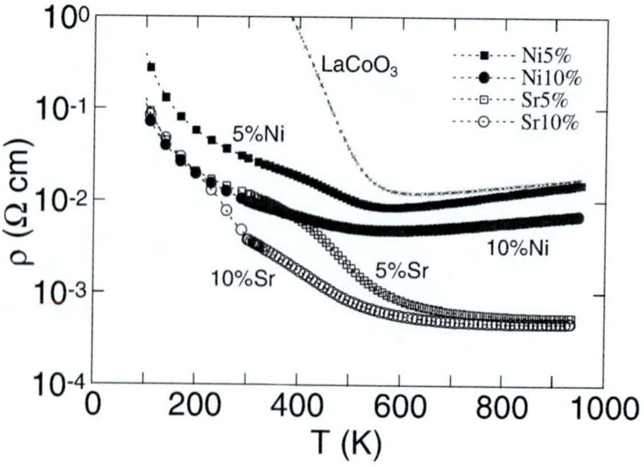

Fig. 1.29 Temperature dependence of ρ of $La_{1-x}Sr_xCoO_3$ and $LaCo_{1-x}Ni_xO_3$ The figure was redrawn using the data in [15]

It has been reported that the electron-doped $LaCoO_3$ ($La_{1-x}Th_xCoO_3$) also shows the IMT at around 500 K [55]. Gerthsen and Härdil reported the significant decrease of the magnitude of TEP along with $\rho(T)$ with increasing temperature at around 500 K for Sr- and Th-doped $LaCoO_3$, which strongly supports the 500 K-transition to be an IMT for both the hole- and electron-doped $LaCoO_3$ [90]. In the electron-doped system $Y_{1-x}Ce_xCoO_3$, the properties of the 500 K-transition such as the transition temperature and the increment of the magnetization seem to be almost independent of the electron-doping by Ce-substitution [68].

To summarize this section, the 500 K-transition is common for $RECoO_3$. Both the spin-state change of Co^{3+} and IMT occur at around 500–700 K, and the substances become magnetic metals at higher temperatures above the 500 K-transition, independent of the RE and/or carrier-doping. These facts imply that the electronic structure unique to $3d^6$ electronic configuration in trivalent perovskite cobaltites is responsible for the 500 K-transition. Only $LaCoO_3$ exceptionally, a part of the spin-state transition might take place at approximately 100 K. The spin-state of Co^{3+} in $RECoO_3$ induced above the 500 K-transition is believed to be HS by most of the researchers, although its experimental basis is lacking. The relation between the spin-crossover and IMT at 500–700 K is not obvious, i.e., it is unknown whether both phenomena inevitably originated from the electron-orbital state, or independent phenomena accidentally occur in the same temperature region. Since the spin-crossover and IMT occur simultaneously in the same temperature region unexceptionally in all $RECoO_3$, clarifying the mechanism to couple the spin-crossover and IMT at 500–700 K is important. More investigations featuring the electronic structure of Co-$3d^6$ orbital state below and above the 500 K-transition are indispensable in order to address the problem.

1.6 Recent Progress

1.6.1 Imaging of Electron-Orbital State of Cobalt Oxides Using X-ray Compton Scattering

In this subsection, we introduce a study of the electronic structure of cobaltites responsible for its spin-crossover phenomena not taken up in Chap. 2.

The X-ray Compton scattering experiments enable the imaging of the electron density distribution in the momentum space [91–95]. The Compton profile $J(p_z)$ is the projection of the three-dimensional electron momentum density $\rho(p_x, p_y, p_z)$ onto the z axis, which lies along the scattering vector, as

$$J(p_z) = \iint \rho(p_x, p_y, p_z) dp_x dp_y. \tag{1.1}$$

The electron momentum density (EMD) can be reconstructed from $J(p_z)$ measured along the several crystallographic directions. Comparison of EMDs below and above the spin-crossover of $LaCoO_3$ demonstrates the Co-3d electron-orbital states responsible for its spin-crossover phenomena [96, 97].

The Compton scattering experiments were conducted at 10 and 270 K in seven directions between $[100]_c$ and $[100]_c$ of the pseudo-cubic cell of $LaCoO_3$. The difference of the Compton profiles between 10 and 270 K ($J_{10K} - J_{270K}$) clarified the symmetry change in the EMD of Co-3d orbitals across the 100 K-transition as shown in Fig. 1.30. The two-dimensional difference EMD (2D-EMD) shown in Fig. 1.31 was reconstructed from the difference Compton profiles. As shown in Fig. 1.31a, b, a positive component of the 2D-EMD extends toward the outer regions of $\langle 110 \rangle_c$, whereas a negative component is observed for $\langle 100 \rangle_c$. The symmetry of the 2D-EMD reflects that of the electron density in the real space. Therefore, these findings provide a direct evidence for the electron transfer from t_{2g} (xy, yz, and zx) to e_g ($3z^2 - r^2$ and $x^2 - y^2$) orbitals through the 100 K-transition [96]. The Compton scattering experiments were also conducted across the 500 K-transition, and clarified the similar difference Compton profiles and the reconstructed 2D-EMD [97]. The Compton scattering experiments unambiguously demonstrate that the electron transfer between t_{2g} and e_g orbitals occurs both at the 100 K- and 500 K-transitions [96, 97].

As can be easily understood from (1.1), the temperature dependence of the area intensity of the difference Compton profile $I_{\Delta J}(T) = \int_0^{p_z^{max}} |J_{10K} - J_T| dp_z$ (J_T is the Compton profile at measured temperature T, and p_z^{max} = 10 a.u. in the present experiment) reflects the magnitude of the symmetry change of the EMD of Co-3d owing to the t_{2g}–e_g electron transfer. The $I_{\Delta J}(T)$ for $[100]_c$ (see Fig. 1.32) shows a thermal-activation-like increment with increasing temperature below 300 K. On the other hand, a steep increment was observed at around 500 K as shown in Fig. 1.32, which suggests the cooperative character of the 500 K-transition.

The shape of the Compton profiles and 2D-EMD also provide an important information relevant to the electron-orbital state responsible for the spin-crossover phenomena in $LaCoO_3$. The difference Compton profile for $[100]_c$ shows an oscillation at approximately p_z = 2–4 atomic units (a.u.), which mainly arises from J_{270K} (see Fig. 1.30). The oscillation is characteristic of the molecular orbitals (MO) constructed by hybridization between Co-e_g and O-2p orbitals [92–95]. In fact, the oscillatory behavior is well reproduced in the difference Compton profiles calculated for the MO, whereas not for localized atomic orbitals (Atom) as shown in Fig. 1.30. The experimentally obtained 2D-EMD has appreciable structure in the region with p_x and p_y less than ~2 a.u. as shown in Fig. 1.31a and 1.31b. This characteristic is in agreement with that for the calculated 2D-EMD of MO taking into account the Co(3d)-O(2p) hybridization [see Fig. 1.31c, d], and much different from that of Atom [see Fig. 1.31e, f]. The oscillatory behavior in the difference Compton profile, which is evolved by the 100 K-transition, develops with increasing temperature and becomes much prominent above 500 K as shown in Fig. 1.33. This finding exhibits a further development of the MO formation resulting from the progress of

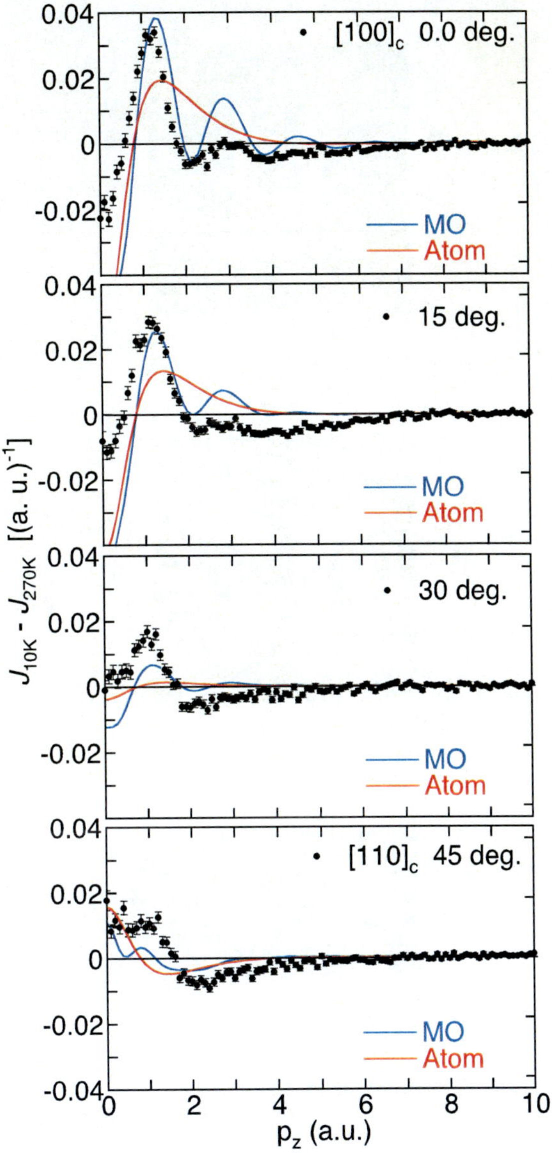

Fig. 1.30 Difference Compton profiles ($J_{10K} - J_{270K}$) of $LaCoO_3$, along with those calculated for molecular (MO) and atomic (Atom) orbitals [96]. (Reprinted from [96], Copyright (2015) The Physical Society of Japan.)

the hybridization between Co-e_g and O-2p orbitals through the 500 K-transition, which might be related to the enhanced itineracy of the electrons above IMT at approximately 500 K.

The measured Compton profiles imply that the MO resulting from the covalent bonding between Co-3d and O-2p orbitals plays an important role in the spin-crossover phenomena of $LaCoO_3$. This fact strongly suggests the collective character

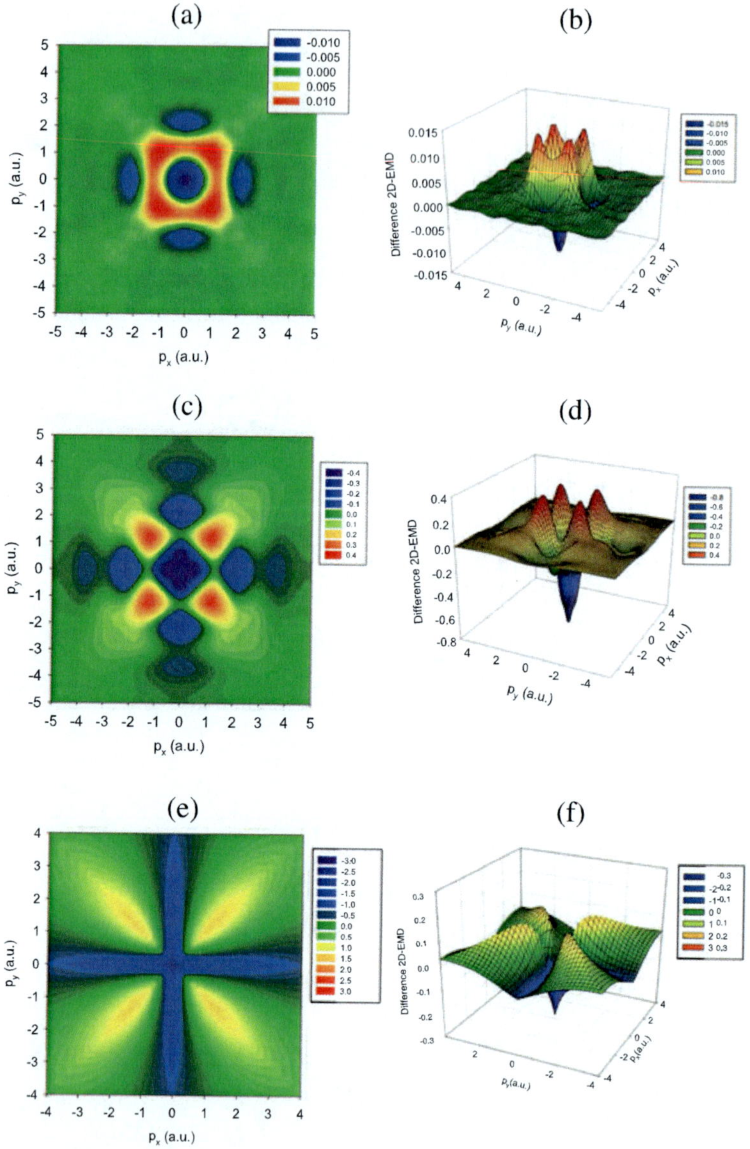

Fig. 1.31 **a** Aerial view and **b** contour plot of the difference two-dimensional electron momentum density (2D-EMD) of Co-3*d* in $LaCoO_3$ between 10 and 270 K, along with those of the calculated molecular orbitals (MO) [**c** and **d**] and localized atomic orbitals (Atom) [**e** and **f**] [96]. (Reprinted from [96], Copyright (2015) The Physical Society of Japan.)

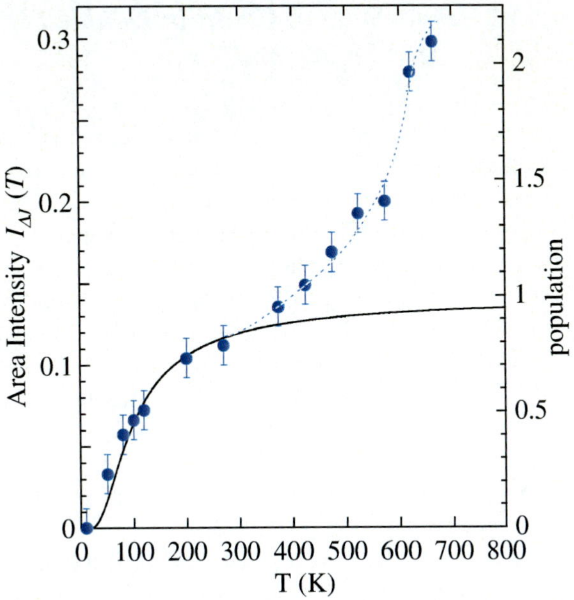

Fig. 1.32 Temperature dependence of the area intensity of the difference Compton profiles ($I_{\Delta J}(T)$). The figure was redrawn using the data in [97]

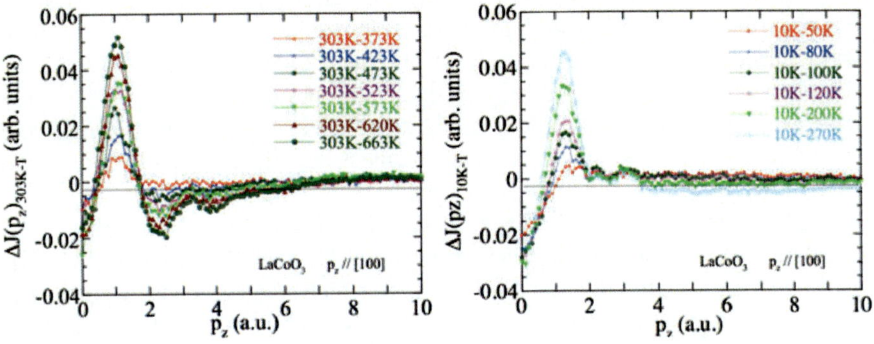

Fig. 1.33 Temperature dependence of the difference Compton profiles [96, 97]

of the spin-crossover in $LaCoO_3$, i.e., the contribution of nonlocal electron-orbital state is important for the spin-crossover phenomena.

As described above, the X-ray Compton scattering experiments on $LaCoO_3$ have demonstrated the images of the electron-momentum-density distributions. The Compton profiles and 2D-EMD revealed the characteristics of the Co-3*d* orbital state, which provides insights unobtainable by other electronic structure experiments.

1.6.2 Spin-State of Co-3d in Double Perovskites Sr_2CoMO_6 (M = Sb, Nb and Ta)

As described in Sect. 1.1, the ligand-field theory predicts that HS ground state is realized for $3d^6$ when the intraatomic exchange interaction surpasses the crystalline-field splitting, i.e., the region for $\Delta_0/B < 20$ in Fig. 1.8. There are quite a few reports on perovskite $3d^6$ cobaltites in which Co-3d electrons show HS ground state, in contrast to the large number of those with LS ground state. Therefore, it is significantly meaningful to synthesize Co^{3+} oxides with the HS ground state, and to investigate the physical properties of those substances. In the double perovskite Sr_2CoMO_6 where M represents pentavalent ions such as Sb^{5+}, Nb^{5+}, and Ta^{5+}, the HS ground state is expected owing to the small Δ_0 at the Co^{3+}-site resulting from the elongation of Co-O distance since M^{5+} attracts O^{2-} ions. In fact for Sr_2CoSbO_6 with almost no oxygen deficiency, the HS state with p_{eff} = ~5 μ_B/Co has been reported [98–100]. For M = Nb and Ta, however, the magnetic properties have not been clear.

Kobayashi et al. investigated the crystalline structure and the magnetic properties of Sr_2CoMO_6 with M = Sb, Nb, and Ta [101]. According to the intensity of the superlattice reflections in the powder X-ray diffraction patterns for the pseudo-cubic unit cells, they clarified that an almost completely ordered double perovskite structure was realized for M = Sb [see Fig. 1.34a]. On the other hand for M = Nb and Ta, the superlattice reflections cannot be seen, indicating the ionic disorder of Co and M ions [see Fig. 1.34b, c]. The difference between the ionic order and disorder due to the difference of M^{5+} cannot be attributed to the gain of the Madelung energy arising from the charge difference between M^{5+} and Co^{3+}. Temperature dependence of χ of Sr_2CoMO_6 are summarized in Fig. 1.35. For M = Sb, the Curie–Weiss-like behavior of $\chi(T)$ is obvious above the Neel temperature T_N =~40 K [see Fig. 1.35a]. The p_{eff} obtained from the fitting of $\chi(T)$ is 3.38, which suggests that Co^{3+} ions are in either magnetic HS or IS ground state. For M = Nb and Ta, Curie–Weiss behavior in $\chi(T)$ is unclear [see Fig. 1.35b and c]. p_{eff} obtained from the fitting is much smaller than that for M = Sb (p_{eff} = 1.91 and 1.79 for Nb and Ta, respectively), which implies that the materials at least partly contain LS Co^{3+}. It was clarified that the introduction of Co^{2+} by the La^{3+}-substitution into Sr-site promotes the ionic order between Co^{2+} and M^{5+}, and yields the HS state of Co^{2+} [101].

The experimental results described above support that the reduction of Δ_0 at Co-site due to Co-M^{5+} ionic order promotes HS Co. However, the difference of the ionic order/disorder between the oxides with M = Sb, and M = Nb and Ta cannot be derived from the localized Co-3d state under the crystalline-electric field.

In order to explain the difference, a model taking into account the covalency among Co-O-M should be necessary. Since the Pauling electronegativity of Sb (2.1) is considerably larger than those of Nb (1.6) and Ta (1.5) [103], Sb^{5+}-O^{2-} bonding is expected to be more covalent than Nb^{5+}-O^{2-} and Ta^{5+}-O^{2-}. Thus, the valence orbitals of Sb hybridize largely with O-2p orbitals, and the occupied O-2p electrons are attracted to Sb^{5+}. This mechanism stabilizes Co^{3+}-O^{2-}-Sb^{5+} arrangement com-

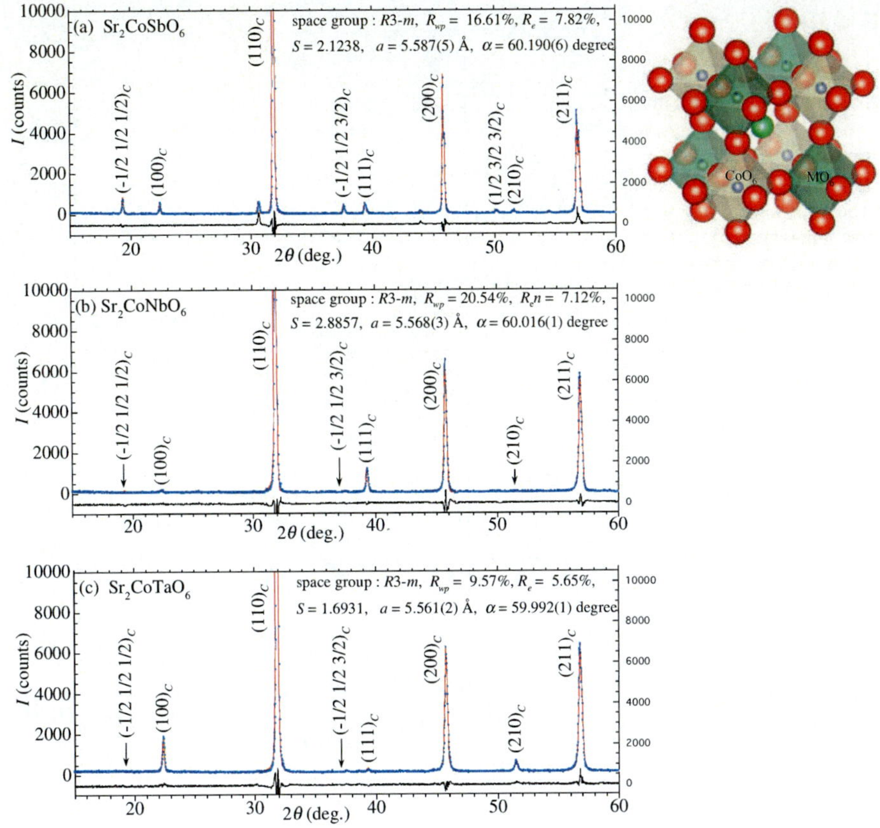

Fig. 1.34 Powder X-ray diffraction patters of Sr_2CoMO_6 with **a** M = Sb, **b** Nb, and **c** Ta [101], along with the structure model for the ordered double perovskite Sr_2CoMO_6 [6]. The figures were redrawn using the data in [101]. The points show the measured profiles and the solid lines show fitted ones by the Rietveld method and the difference between measured and fitted data [102]

pared with Sb^{5+}-O^{2-}-Sb^{5+} one, leading to the ionic order of Co^{3+} and Sb^{5+}. The hybridization between the orbitals of Sb and O is expected to reduce Δ_0 at Co-site, which promotes HS Co. In the cases of M = Nb and Ta, the covalent characters of M^{5+}-O^{2-} and that of Co^{3+}-O^{2-} are not much different, which results in the ionic disorder of Co^{3+} and M^{5+}. The larger covalent bonding among Sb-5s, 5p, and O-2p compared with that among Nb-5s, 5p or Ta-5s, 5p, and O-2p has been clarified by the hard X-ray photoemission spectroscopy (HAXPES) experiments. As the number of Co^{2+} increases owing to the introduction of La^{3+}, Co^{2+}-O^{2-}-M^{5+} arrangement develops resulting from the larger gain of the Madelung energy by the ionic order between Co and M ions. In fact, $(Sr_{1-x}La_x)_2CoMO_6$ (M = Nb and Ta) shows a fairly large degree of the ionic order for $x \geq 0.3$ [101].

According to the study described above, Co-M ionic order/disorder and the spin-states of Co ions are strongly correlated, i.e., the ionic order promotes the HS state

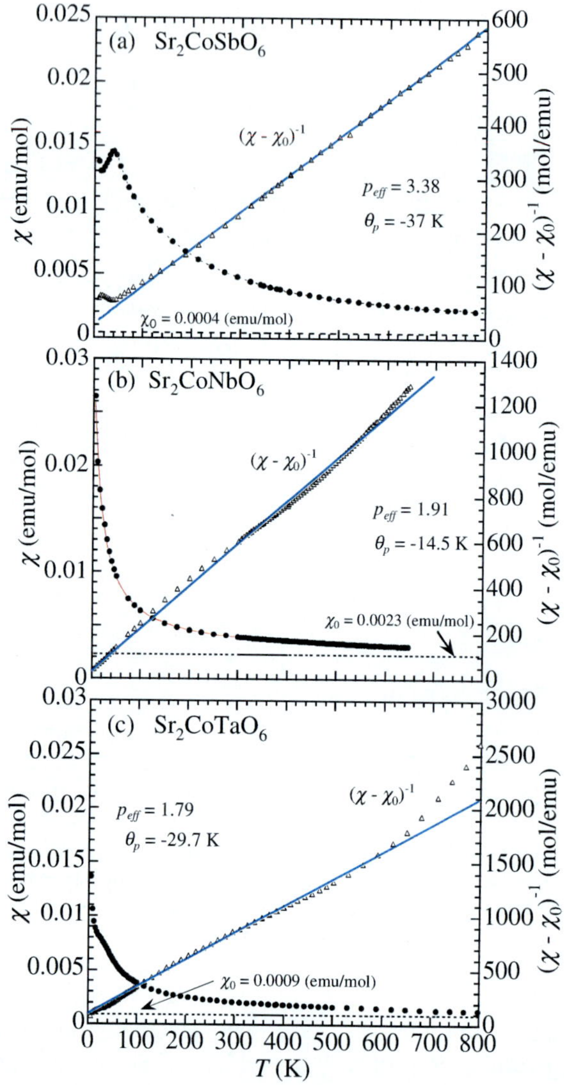

Fig. 1.35 $\chi(T)$ and $(\chi - \chi_0)^{-1}(T)$ of Sr_2CoMO_6 with M = Sb, Nb, and Ta. The solid lines represent the fitted ones with temperature-independent (χ_0) and Curie–Weiss terms. The figures were redrawn using the data in [101]

of Co ions. This fact is in agreement with the expectation that the HS state will be the ground state for 3d electrons when Δ_0 is smaller than the intraatomic exchange interaction. On the other hand, the difference between M = Sb, and M = Nb and Ta implies that the competition of the covalency between Co-3d and O-2p, and O-2p and M-5s, 5p has an important role to determine the electronic-spin state of Co-3d. This perspective also supports that the electron-orbital state of Co is not the local ionic 3d state and the covalency or MO character should be taken into account to evaluate the magnetic properties such as the spin-crossover phenomena.

1.6.3 Hybridization Effect Between Co-3*d* and Pr-4*f* via O-2*p* in $Pr_{1-x}Y_xCoO_3$

Although the studies on $RECoO_3$ have been actively carried out, the research interest has mainly been focused on the magnetic properties of Co-3*d* electrons. On the other hand, the magnetism of the RE-4*f* electrons has not aroused the widespread interest of researchers, since it has been considered that the RE-4*f* are well-localized and the RE-4*f* magnetism is independent of the Co-3*d* magnetism. In this subsection, however, we introduce an interesting example where RE-4*f* and Co-3*d* magnetically correlate to each other.

Figure 1.36 shows the temperature dependence of χ of perovskite $La_{1-x}Pr_xAlO_3$ with $x = 0$, 0.5, and 1 [89]. Since both Al and La are nonmagnetic, $\chi(T)$ of $La_{1-x}Pr_xAlO_3$ certainly reflects the magnetism of Pr-4*f* electrons. $\chi(T)$, which does not follow the ordinary Curie–Weiss law as shown in Fig. 1.36, can be well reproduced by the fitting based on the crystalline-electric-field model with the nonmagnetic singlet Γ_1 ground state, the first-excited nonmagnetic doublet Γ_3, and the second-excited magnetic triplet Γ_4, i.e., $\chi(T)$ exhibits the temperature-dependent Van Vleck paramagnetism with the finite values at the lowest temperature [104]. The result suggests that the Pr^{3+} with $4f^2$ in perovskite structure usually have a nonmagnetic ground state Γ_1 for well-localized 4*f*-orbitals. $\chi(T)$ of $PrCoO_3$ shown in Fig. 1.37 does not show a tendency to diverge, similar to that of $PrAlO_3$, although $\chi(T)$ of $PrCoO_3$ cannot be reproduced by the crystalline-electric-field model with any set of energy levels [89]. The difference of $\chi(T)$ between that of $PrCoO_3$ and $PrAlO_3$ [$\Delta\chi(T)$] is also shown in Fig. 1.37. A gradual increment of $\Delta\chi(T)$ with increasing temperature above 200 K is possibly due to the spin-crossover of Co^{3+} [89] as described in Sect. 1.5. $\Delta\chi(T)$ arising from Pr-4*f* is apparent below 100 K. This fact implies that the Pr-4*f* ground state is somewhat different from the nonmagnetic singlet Γ_1 for the localized $4f^2$. Saitoh et al. have reported a preliminary study of the resonant photoemission spectra of $PrCoO_3$, suggesting a deviation from the ordinary Pr-$4f^2$ state [105].

In order to clarify the detailed magnetic properties of Pr-4*f*, it is useful to investigate Pr-cobaltites in which the Pr-site is diluted with nonmagnetic trivalent ions. However, $La_{1-x}Pr_xCoO_3$ is unsuitable due to the 100 K-transition of Co-3*d* for lower *x* (see Sect. 1.5). Kobayashi et al. investigated $Pr_{1-x}Y_xCoO_3$ crystal, where the measured magnetization at lower temperatures reflects only the Pr-4*f* magnetism because of the LS state of Co^{3+} ions below at around 500 K for all Y contents *x* [106].

For the dense Pr content, a non-Curie–Weiss-like $\chi(T)$ similar to that of $PrCoO_3$ is observed. Most strikingly, the Curie–Weiss-like $\chi(T)$, which tends to diverge at 0 K, becomes prominent as the decrease of Pr content by Y-substitution [see Fig. 1.38a]. The result indicates that the magnetic ground state is realized for Pr-4*f* in $Pr_{1-x}Y_xCoO_3$ with dilute Pr content. The values of p_{eff} per Pr ion obtained from the Curie–Weiss fitting of the measured $\chi(T)$ are closer to the value predicted by the Hund rule for the $4f^1$ electronic configuration rather than that for $4f^2$ [106]. It should be noted that the change of the 4*f* ground state from nonmagnetic to mag-

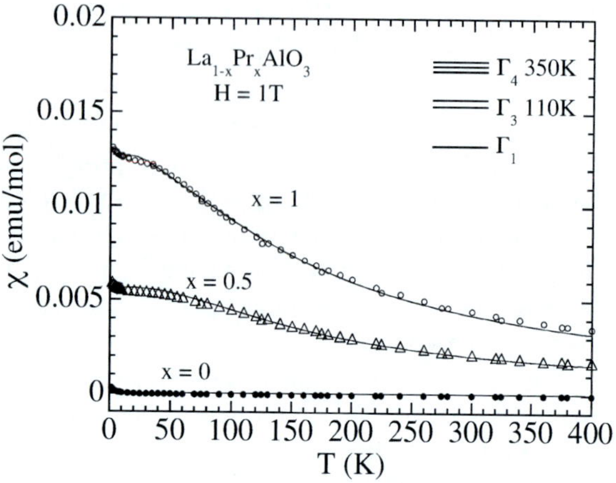

Fig. 1.36 Temperature dependence of χ of $La_{1-x}Pr_xAlO_3$ ($x = 0, 0.5$, and 1). The solid lines show the calculated $\chi(T)$ based on the crystalline-electric-field model with the energy levels shown in the inset [89]. (Reprinted from [89], Copyright (2006) The Physical Society of Japan.)

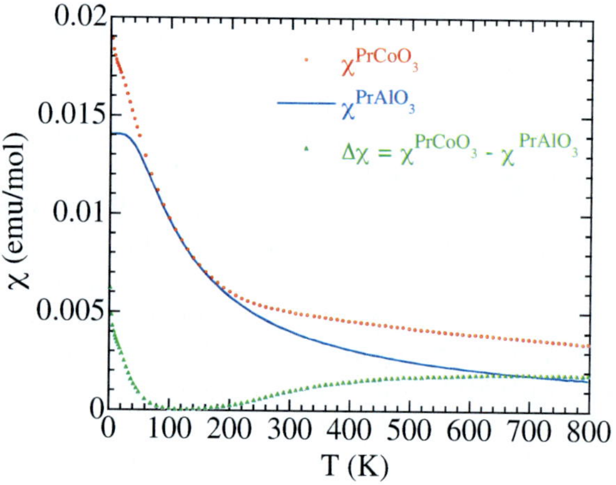

Fig. 1.37 $\chi(T)$ of $PrCoO_3$ and $PrAlO_3$, and their difference $\Delta\chi$ [89]. (Reprinted from [89], Copyright (2006) The Physical Society of Japan.)

netic takes place only for the combination of Pr and Co. In contrast to $\chi(T)$ of Pr in $PrCoO_3$, $\chi(T)$ of Pr in $PrAlO_3$ and $\chi(T)$ of Eu in $EuCoO_3$ do not change with the introduction of Y ions [see Fig. 1.38b, c].

The change of the Pr-4f ground state with decreasing Pr content from nonmagnetic to magnetic suggests that the $4f^2$ state for dense Pr changes to the $4f^1$ magnetic ground state for dilute Pr. The resonant photoemission spectra of Pr in $PrCoO_3$ have revealed a characteristic double peak structure similar to that of $PrBa_2Cu_3O_7$, where the Pr^{4+} ($4f^1$) is realized by ligand hole transfer from CuO_2 to Pr owing to the hybridization between Pr-4f and Cu-3d via O-2p [105]. According to this fact, a similar mechanism is expected to cause the changes of the Pr-4f magnetism of $Pr_{1-x}Y_xCoO$.

In order to explain the Pr-4f magnetism of $Pr_{1-x}Y_xCoO$, a simple model based on the ligand hole transfer among the hybridized Co-3d and Pr-4f via O-2p states is

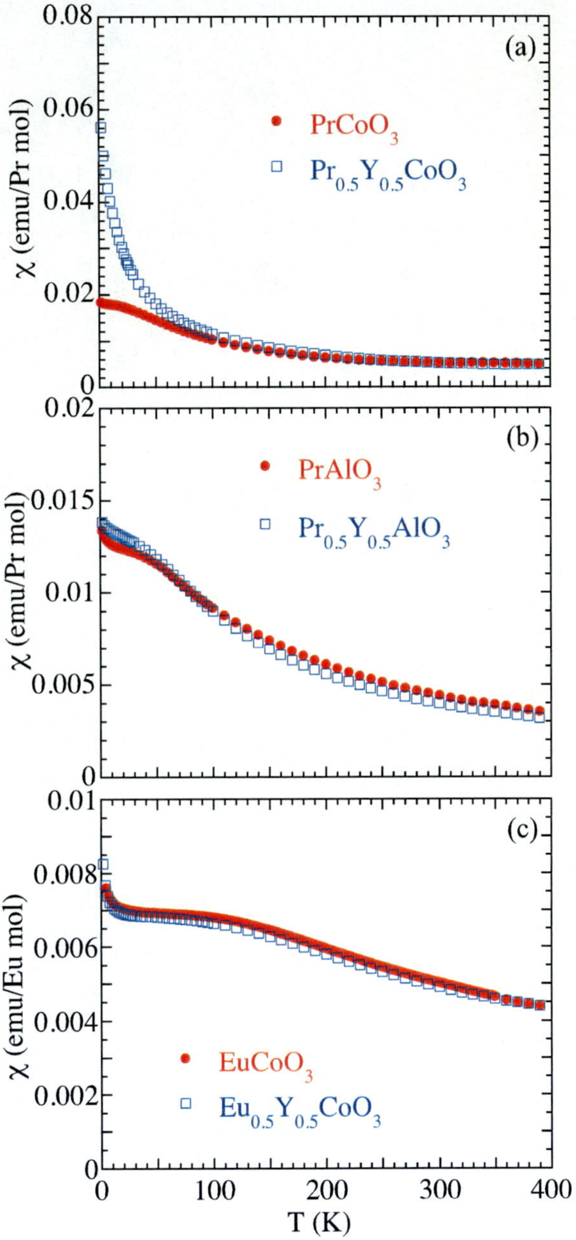

Fig. 1.38 Temperature dependence of χ per Pr or Eu mol for **a** $PrCoO_3$ and $Pr_{0.5}Y_{0.5}CoO_3$, **b** $PrAlO_3$ and $Pr_{0.5}Y_{0.5}AlO_3$, and **c** $EuCoO_3$ and $Eu_{0.5}Y_{0.5}CoO_3$.[106]. (Reprinted from [106], Copyright (2014) The Physical Society of Japan.)

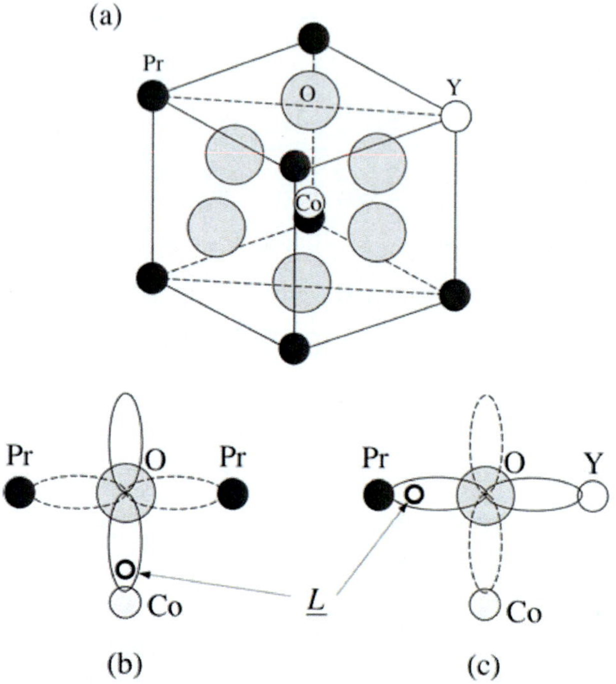

Fig. 1.39 **a** Pseudo-cubic unit cell of $Pr_{1-x}Y_xCoO_3$. Schematic figures showing the bonding among Pr, Y, O, and Co in the $\{110\}_c$ plane of the pseudo-cubic unit cell. Ligand hole transfer for the case that **b** all RE-sites are occupied by Pr ions and **c** Y ions are partially introduced, in which $\underline{L}$ denotes a ligand hole [106]. (Reprinted from [106], Copyright (2014) The Physical Society of Japan.)

proposed (see Fig. 1.39). In the pseudo-cubic unit cell of $Pr_{1-x}Y_xCoO_3$, RE-O–Co bondings are realized in the $\{110\}_c$ plane of the pseudo-cubic unit cell.

It is expected that the Co-3d ($|3d^6\rangle$) and O-2p ($|2p^6\rangle$) states form molecular orbitals through covalent bonds [2, 18], and create ligand holes $\underline{L}$ as $|3d^7\underline{L}\rangle$. When the hybridization between Pr-4f ($|4f^2\rangle$) and O-2p also exists, ligand hole transfer from Co(3d)-O(2p) molecular orbitals to the Pr-4f becomes possible. In $PrCoO_3$, where all the RE-sites are occupied by Pr ions, each Pr-site is equivalent from the O-sites. In this case, the transfer of $\underline{L}$ to a specific Pr-site does not occur and the ligand holes $\underline{L}$ stay in Co(3d)-O(2p) molecular orbitals as $|3d^7\underline{L}\rangle$. Thus, Pr-4$f$ states maintain nonmagnetic $|4f^2\rangle$ [see Fig. 1.39b]. The introduction of Y ions into RE-sites decreases the number of Pr ions, which increases the probability of the ligand hole transfer to Pr-4f since Y ions have no 4f electrons ($|4f^0\rangle$) [see Fig. 1.39c]. The ligand hole transfer causes the electronic state of Pr to become $|4f^2\underline{L}\rangle$, which is equivalent to the $|4f^1\rangle$ ($^2F_{5/2}$) state, leading to a magnetic ground state. The formation of $|4f^2\underline{L}\rangle$ states naturally becomes marked for larger Y contents. Note that even for non-doped $PrCoO_3$, the $|4f^2\rangle$ state of Pr is different from that of completely localized Pr such as in $PrAlO_3$, since Pr-4f states hybridize with Co(3d)-O(2p) molecular orbitals.

Although the topic described above is not an issue of the spin-crossover of Co-3*d*, it is an example of collective phenomena in which molecular orbitals formed by covalent bondings among Co-3*d* and the orbitals of other elements play an important role. The investigation of the electronic structure and the ligand hole transfer model using photoemission spectroscopy experiments will be described in Chap. 2 by Saitoh.

References

1. J.B. Goodenough, J.M. Longo, *Landort-Börunstein vol. 4a, chap. 3* (1970)
2. M. Imada, A. Fujimori, Y. Tokura, Rev. Mod. Phys. **70**, 1039 (1998)
3. A.K. Kundu (ed.), *Magnetic Perovskites - Synthesis, Structure and Physical Properties* (Springer, 2016)
4. Bussmann-Holder, A.K. Hugo (eds.), *High Superconductors and Related Transition Metal Oxides* (Springer-Verlag Berlin Heidelberg, 2007)
5. J.B. Goodenough, J. Phys. Chem. Solids **6**, 287 (1958)
6. K. Momma, F. Izumi, J. Appl. Crystallogr. **44**, 1272 (2011)
7. Y. Tanabe, S. Sugano, J. Phys. Soc. Jpn. **9**, 753 (1954)
8. R.D. Shannon, Acta Crystallogr., Sect. A **32**, 751 (1976)
9. Y. Tanabe, S. Sugano, J. Phys. Soc. Jpn. **9**, 766 (1954)
10. R.H. Potze, G.A. Sawatzky, M. Abbate, Phys. Rev. B **51**, 11501 (1995)
11. M.A. Korotin, S.Y. Ezhov, I.V. Solovyev, V.I. Anisimov, D.I. Khomskii, G.A. Sawatzky, Phys. Rev. B **54**, 5309 (1996)
12. T. Mizokawa, A. Fujimori, Phys. Rev. B **54**, 5368 (1996)
13. R.R. Heikes, R.C. Miller, R. Mazelsky, Physica **30**, 1600 (1964)
14. K. Asai, A. Yoneda, O. Yokokura, J.M. Tranquada, G. Shirane, K. Kohn, J. Phys. Soc. Jpn. **67**, 290 (1998)
15. Y. Kobayashi, S. Murata, K. Asai, J.M. Tranquada, G. Shirane, K. Kohn, J. Phys. Soc. Jpn. **68**, 1011 (1999)
16. W.C. Koehler, E.O. Wollan, J. Phys. Chem. Solids **2**, 100 (1957)
17. K. Asai, O. Yokokura, N. Nishimori, H. Chou, J.M. Tranquada, G. Shirane, S. Higuchi, Y. Okajima, K. Kohn, Phys. Rev. B **50**, 3025 (1994)
18. T. Saitoh, T. Mizokawa, A. Fujimori, M. Abbate, Y. Takeda, M. Takano, Phys. Rev. B **55**, 4257 (1997)
19. J. Kanamori, J. Phys. Chem. Solids. **10**, 87 (1959)
20. Y. Kobayashi, N. Fujiwara, S. Murata, K. Asai, H. Yasuoka, Phys. Rev. B **62**, 410 (2000)
21. S. Murata, S. Ishida, M. Suzuki, Y. Kobayashi, K. Asai, K. Kohn, Physica B **263–264**, 647 (1999)
22. S. Yamaguchi, Y. Okimoto, Y. Tokura, Phys. Rev. B **55**, R8666 (1997)
23. G. Maris, Y. Ren, V. Volotchaev, C. Zobel, T. Lorenz, T.T.M. Palstra, Phys. Rev. B **67**, 224423 (2003)
24. A. Ishikawa, J. Nohara, S. Sugai, Phys. Rev. Lett **93**, 136401 (2004)
25. Y. Kobayashi, T.S. Naing, M. Suzuki, M. Akimitsu, K. Asai, K. Yamada, J. Akimitsu, P. Manuel, J.M. Tranquada, G. Shirane, Phys. Rev. B **72**, 174405 (2005)
26. T.S. Naing, T. Kobayashi, Y. Kobayashi, M. Suzuki, K. Asai, J. Phys. Soc. Jpn. **75**, 084601 (2006)
27. D. Louca, J.L. Sarrao, Phys. Rev. Lett **91**, 155501 (2003)
28. D. Phelan, D. Louca, S. Rozenkranz, S.H. Lee, Y. Qiu, P.J. Chupas, R. Osorn, H. Zeng, J.F. Mitchell, J.R.D. Copley, J.L. Sarrao, Y. Moritomo, Phys. Rev. Lett **96**, 027201 (2006)
29. J. Kanamori, J. Appl. Phys. **31**, 14S (1960)

30. J. Goodenough, Phys. Rev. **100**, 564 (1955)
31. Y. Murakami, J.P. Hill, D. Gibbs, M. Blume, I. Koyama, M. Tanaka, H. Kawata, T. Arima, Y.T.K. Hirota, Y. Endoh, Phys. Rev. Lett. **81**, 582 (1998)
32. R.J. Radwański, Z. Ropka, Solid State Commun. **112**, 621 (1999)
33. R.J. Radwański, Z. Ropka, Physica B. **281–282**, 507 (2000)
34. S. Noguchi, S. Kawamata, K. Okuda, H. Nojiri, M. Motokawa, Phys. Rev. B **66**, 094404 (2002)
35. A. Podlesnyak, S. Streule, J. Mesot, M. Medarde, E. Pomjakushina, K. Conder, A. Tanaka, M.W. Haverkort, D.I. Khomskii, Phys. Rev. Lett. **97**, 247208 (2006)
36. Z. Ropka, R.J. Radwanski, Physica B. **312–313**, 777 (2002)
37. Z. Ropka, R.J. Radwanski, Phys. Rev. B. **67**, 172401 (2003)
38. T. Kyômen, Y. Asaka, M. Itoh, Phys. Rev. B **67**, 144424 (2003)
39. T. Kyômen, Y. Asaka, M. Itoh, Phys. Rev. B **71**, 024418 (2005)
40. M.W. Haverkort, Z. Hu, J.C. Cezar, T. Burnus, H. Hartmann, M. Reuther, C. Zobel, T. Lorenz, A. Tanaka, N.B. Brookes, H.H. Hsieh, H.J. Lin, C.T. Chen, L.H. Tjeng, Phys. Rev. Lett. **97**, 176405 (2006)
41. K. Knizek, Z. Jirak, J. Hajmanek, P. Novak, W. Ku, Phys. Rev. B **79**, 014430 (2009)
42. K. Tomiyasu, J. Okamoto, H.Y. Huang, Z.Y. Chen, E.P. Sinaga, W.B. Wu, Y.Y. Chu, A. Singh, R.P. Wang, F.M.F. de Groot, A. Chainani, S. Ishihara, C.T. Chen, D.J. Huang, Phys. Rev. Lett. **119**, 196402 (2017)
43. K. Sato, A. Matsuo, K. Kindo, Y. Kobayashi, K. Asai, J. Phys. Soc. Jpn. **78**, 093702 (2009)
44. M.M. Altarawneh, G.W. Chern, N. Harrison, C.D. Batista, A. Uchida, M. Jaime, D.G. Rickel, S.A. Crooker, C.H. Mielke, J.B. Betts, J.F. Mitchell, M.J.R. Hoch, Phys. Rev. Lett **109**, 037201 (2012)
45. A. Ikeda, T. Nomura, Y.H. Matsuda, A. Matsuo, K. Kindo, K. Sato, Phys. Rev. B **93**, 220401(R) (2016)
46. T. Tatsuno, E. Mizoguchi, J. Nasu, M. Naka, S. Ishihara, J. Phys. Soc. Jpn. **85**, 083706 (2016)
47. M. Itoh, M. Sugahara, I. Natori, K. Motoya, J. Phys. Soc. Jpn. **64**, 3967 (1995)
48. M. Itoh, J. Hashimoto, S. Yamaguchi, Y. Tokura, Physica B **281–282**, 510 (2000)
49. Y. Shimizu, T. Takahashi, S. Yamada, A. Shimokata, T. Jin-no, M. Itoh, Phys. Rev. Lett. **119**, 267203 (2017)
50. A. Hariki, A. Yamanaka, T. Uozumi, J. Phys. Soc. Jpn. **84**, 073706 (2015)
51. K. Tomiyasu, T. Nomura, Y. Kobayashi, Ishihara, Ohira-Kawamura, M. Kofu, cond-mat.str-el arXiv:1808.05888vl (2018)
52. P.M. Raccah, J.B. Goodenough, J. Appl. Phys. **39**, 1209 (1968)
53. M. Itoh, I. Natori, S. Kubota, K. Motoya, J. Phys. Soc. Jpn. **63**, 1486 (1994)
54. T. Saitoh, T. Mizokawa, A. Fujimori, M. Abbate, Y. Takeda, M. Takano, Phys. Rev. B **56**, 1290 (1997)
55. V.P. Gerthsen, K.H. Härdil, Z. Naturforsch. **17a**, 514 (1962)
56. S. Yamaguchi, H. Taniguchi, H. Takagi, T. Arima, Y. Tokura, J. Phys. Soc. Jpn. **64**, 1885 (1995)
57. K. Muta, Y. Kobayashi, K. Asai, J. Phys. Soc. Jpn. **71**, 2784 (2002)
58. P. Mandal, Choudhury, K. Biswas, B. Ghosh, Phys. Rev. B **70**, 104407 (2004)
59. M. Kriener, C. Zobel, A. Reichl, J. Baier, M. Cwik, K. Berggold, H. Kierspel, O. Zabara, A. Freimuth, T. Lorenz, Phys. Rev. B **69**, 094417 (2004)
60. M. Kriener, M. Braden, Kierspel, Senff, O. Zabara, C. Zobel, T. Lorenz, Phys. Rev. B **79**, 224104 (2009)
61. H. Masuda, T. Fujita, T. Miyashita, M. Soda, Y. Yasui, Y. Kobayashi, M. Sato, J. Phys. Soc. Jpn **72**, 873 (2003)
62. J. Blasco, J. Garcfa, Phys. Rev. B **51**, 3569 (1995)
63. J. Perez, J. Garcfa, J. Blasco, J. Stankiewicz, Phys. Rev. Lett. **80**, 2401 (1962)
64. M.R. Ibarra, R. Mahendiran, C. Marquina, B. Garća-Landa, J. Blasco, Phys. Rev. B **57**, R3217 (1998)

65. K. Sato, M.I. Bartashevich, T. Goto, Y. Kobayashi, M. Suzuki, K. Asai, A. Matsuo, K. Kindo, J. Phys. Soc. Jpn. **77**, 024601 (2008)
66. T. Kyômen, Y. Murachi, M. Itoh, M. Hanaya, Chem. Mater. **20**, 5114 (2008)
67. D. Fuchs, P. Schweiss, P. Adelmann, T. Schwarz, R. Schneider, Phys. Rev. B **72**, 014466 (2005)
68. Y. Kobayashi, T. Koike, M. Okawa, R. Takayanagi, S. Takei, M. Minohara, M. Kobayashi, K. Horiba, H. Kumigashira, A. Yasui, E. Ikenaga, T. Saitoh, K. Asai, J. Phys. Soc. Jpn. **85**, 114704 (2016)
69. K. Tomiyasu, S. Koyama, M. Watahiki, M. Sato, K. Nishihara, Y. Takahashi, M. Onodera, K. Iwasa, T. Nojima, H. Nojiri, J. Okamoto, D.J. Huang, Y. Yamasaki, H. Nakao, Y. Murakami, J. Phys. Soc. Jpn. **85**, 094702 (2016)
70. K. Tomiyasu, M. Sato, S. Koyama, T. Nojima, R. Kajimmoto, S. Ji, K. Iwasa, J. Phys. Soc. Jpn. **86**, 094706 (2017)
71. S. Yamaguchi, Y. Okimoto, H. Taniguchi, Y. Tokura, Phys. Rev. B **53**, R2926 (1996)
72. A. Podlesnyak, K. Conderc, E. Pomjakushinab, A. Mirmelstein, P. Allenspach, D.I. Khomskii, J. Magn. Magn. Mater. **310**, 1552 (2007)
73. A. Podlesnyak, M. Russina, A. Furrer, A. Alfbnsov, E. Vavilova, V. Kataev, B. Buchner, T. Strässle, E. Pomjakushina, K. Conder, D.I. Khomskii, Phys. Rev. Lett. **101**, 247603 (2008)
74. A. Podlesnyak, G. Ehlers, M. Frontzek, A.S. Sefat, A. Furrer, T. Strässle, E. Pomjakushina, K. Conder, F. Demmel, D.I. Khomskii, Phys. Rev. B **83**, 134430 (2011)
75. K. Sato, A. Matsuo, K. Kindo, Y. Kobayashi, K. Asai, J. Phys. Soc. Jpn. **80**, 104702 (2011)
76. K. Sato, A. Matsuo, K. Kindo, Y. Hara, K. Nakaoka, Y. Kobayashi, K. Asai, J. Phys. Soc. Jpn. **83**, 114712 (2014)
77. S. Asai, N. Furuta, Y. Yasui, I. Terasaki, J. Phys. Soc. Jpn. **80**, 104705 (2011)
78. S. Asai, N. Furuta, R. Okazaki, Y. Yasui, I. Terasaki, Phys. Rev. B **86**, 014421 (2012)
79. K. Knizek, J. Hejtmanek, M. Marysko, Z. Jirak, J. Bursik, Phys. Rev B **85**, 134401 (2012)
80. T. Sudayama, H. Nakao, Y. Yamasaki, Y. Murakami, S. Asai, R. Okazaki, Y. Yasui, I. Terasaki, J. Phys. Soc. Jpn. **86**, 094701 (2017)
81. K. Tomiyasu, N. Ito, R. Okazaki, Y. Takahashi, M. Onodera, K. Iwasa, T. Nojima, T. Aoyama, K. Ohgushi, Y. Ishikawa, T. Kamiyama, S. Ohira-Kawamura, M. Kofu, S. Ishihara, Adv. Quantum. Technol. **1**, 1800057 (2018)
82. Y. Tokura, Y. Okimoto, S. Yamaguchi, H. Taniguchi, T. Kimura, H. Takagi, Phys. Rev. B **58**, R1699 (1998)
83. G. Demazeau, M. Pouchard, P. Hagenmuller, J. Solid State Chem. **9**, 202 (1974)
84. G. Thornton, F.C. Morrison, S. Partington, B.C. Tofield, D.E. Williams, J. Phys. C **21**, 2871 (1988)
85. S. Yamaguchi, Y. Okimoto, Y. Tokura, Phys. Rev. B **54**, R11022 (1996)
86. X. Liu, C.T. Prewitt, J. Phys. Chem. Solids **52**, 441 (1991)
87. K. Berggold, M. Kriener, P. Becker, M. Benomar, M. Reuther, C. Zobel, T. Lorenz, Phys. Rev. B **78**, 134402 (2008)
88. K. Knizek, J. Hejtmanek, Z. Jirak, P. Tomes, P. Henry, G. Andre, Phys. Rev. B **79**, 134103 (2009)
89. Y. Kobayashi, T. Mogi, K. Asai, J. Phys. Soc. Jpn. **75**, 104703 (2006)
90. N. Mott, *Metal-Insulator Transitions* (Taylor and Francis Ltd, London, 1990)
91. M.J. Cooper, P.E. Mijnarends, N. Shiotani, N. Sakai, A. Bansil, *X-ray Compton Scattering* (Oxford University Press, Oxford, 2004)
92. A. Koizumi, S. Miyaki, Y. Kakutani, H. Koizumi, N. Hiraoka, K. Makoshi, N. Sakai, K. Hirota, Y. Murakami, Phys. Rev. Lett. **86**, 5589 (2001)
93. A. Koizumi, T. Nagao, N. Sakai, K. Hirota, Y. Murakami, Phys. Rev. B **74**, 012408 (2006)
94. B. Barbiellini, A. Koizumi, P.E. Mijnarends, W. Al-Sawai, H. Lin, T. Nagao, K. Hirota, M. Itou, Y. Sakurai, A. Bansil, Phys. Rev. Lett. **102**, 206402 (2009)
95. Y. Sakurai, M. Itou, B. Barbiellini, P.E. Mijnarends, R.S. Markiewicz, S. Kaprzyk, J.M. Gillet, S. Wakimoto, M. Fujita, S. Basak, Y.J. Wang, W. Al-Sawai, H. Lin, A. Bansil, K. Yamada, Science **332**, 698 (2011)

96. Y. Kobayashi, Y. Sakurai, M. Itou, K. Sato, K. Asai, J. Phys. Soc. Jpn. **85**, 114706 (2015)
97. Y. Kobayashi, Y. Sakurai, N. Tsuji, K. Sato, K. Asai, Phys. Rev. B **98**, 115154 (2018)
98. G. Blasse, J. Appl. Phys. **36**, 879 (1965)
99. G. Blasse, J. Inorg. Nucl. Chem. **27**, 993 (1965)
100. V. Primo-Martin, M. Jansen, J. Solid State Chem. **157**, 76 (2001)
101. Y. Kobayashi, M. Kamogawa, Y. Terakado, K. Asai, J. Phys. Soc. Jpn. **81**, 044708 (2012)
102. F. Izumi, K. Momma, Solid State Phenom. **130**, 15 (2007)
103. L. Pauling, *The Nature of the Chemical Bond* (Cornel University Press, Ithaca, NY, 1960)
104. K.R. Lea, M.J.M. Leask, W.P. Wolf, J. Phys. Chem. Solids. **23**, 1381 (1962)
105. T. Saitoh, D. Ishii, A. Hachimura, M. Hirose, T.S. Naing, Y. Kobayashi, K. Asai, M. Nakatake, M. Higashiguchi, K. Shimada, H. Namatamec, M. Taniguchi, J. Magn. Magn. Matter. **310**, 981 (2007)
106. Y. Kobayashi, Y. Terakado, K. Asai, J. Phys. Soc. Jpn. **83**, 104704 (2014)

Chapter 2
Experimental Electronic Structure of Co Oxides

Tomohiko Saitoh

Abstract Experimental electronic structure of $LaCoO_3$ and other $RCoO_3$ (R = rare earth) in connection with the spin crossover phenomena in $LaCoO_3$ is reviewed. As the experimental probes, we mostly deal with electron spectroscopic methods, namely photoemission spectroscopy and x-ray absorption spectroscopy. Experimental results are compared with ab initio band-structure calculations and configuration-interaction cluster-model calculations. First, we discuss how electronic-structure measurements can contribute to studying spin crossover phenomena, particularly of $LaCoO_3$. Then in Sect. 2.2, we give a brief historical review of electron spectroscopic studies on $LaCoO_3$ in order to highlight the conflict between the two major models, namely the low spin-high spin model and the low spin-intermediate spin-high spin model. Sections 2.3 to 2.5 are devoted to discussing this conflict and how we may be able to resolve it beyond this binarism. In Sect. 2.6, we remark the probing depth of the methods and the surface magnetism of $LaCoO_3$. In Sect. 2.7, we briefly discuss a part of recent advances both in experiment and theory in this field. We run through the electronic structure of $RCoO_3$ and other related Co oxides in Sect. 2.8, and finally we give the concluding remarks in Sect. 2.9.

Keywords Perovskite-type co oxides · Electronic structure · Photoemission spectroscopy · X-ray absorption spectroscopy · Spin crossover

2.1 Introduction

Electron spectroscopies, such as photoemission spectroscopy (PES) and x-ray absorption spectroscopy (XAS), are the best experimental technique to observe the electronic structure of solids. For example, angle-resolved photoemission spectroscopy (ARPES) played a decisive role in determining the symmetry of the superconducting gap in cuprate high-T_C superconductors [1]. However, PES or ARPES

T. Saitoh (✉)
Department of Applied Physics, Tokyo University of Science,
Katsushika, Tokyo 125-8585, Japan
e-mail: t-saitoh@rs.tus.ac.jp

Y. Okimoto et al. (eds.), *Spin-Crossover Cobaltite*, Springer Series
in Materials Science 305, https://doi.org/10.1007/978-981-15-7929-5_2

are in general insensitive to the spin configuration with no magnetic ordering because the photoemission processes do not include any spin interaction [2]. Transition metal (TM) L edge ($2p$ to $3d$) XAS of $3d$ TM compounds has widely been used to determine the spin configuration and the number of electrons in the $3d$ state because the multiplet structure appears in the spectra is sensitive to these quantities. O K edge ($1s$ to $2p$) XAS of TM oxides roughly represents the local density of states of the unoccupied states at the oxygen atom. It can be utilized as an unoccupied counterpart of valence-band PES and is usually insensitive to the spin configuration like PES.

From the above overview, one may wonder why PES has been used from the early stage of the research history of the temperature-induced spin crossover problem in $LaCoO_3$ and why XAS appeared later, as described below in Sect. 2.2. The reason for that is simple but profound; it is all owing to the very unique electronic structure of Co^{3+} ion under the octahedral symmetry in perovskite-type oxides.

Under the local octahedral (O_h) symmetry, the degenerated five $3d$ atomic orbitals split into the three lower t_{2g} orbitals and the two upper e_g orbitals. The spin configuration is determined by the competition between the energy gain due to the Hund's rule coupling and the energy cost for placing electrons on the e_g orbitals over the crystal field splitting ($10Dq$). It is noted that the two-electron exchange interaction, which is the origin of the Hund's rule coupling, is almost constant over the series of the $3d$ elements, whereas $10Dq$, which is actually governed by the hybridization between the TM $3d$ and the ligand O $2p$ orbitals, gradually increases with the TM atomic number. As a result, the Hund's rule coupling wins against $10Dq$ in the early and middle $3d$ TM (Ti-Fe) oxides and the TM ion takes the high spin (HS) configuration while in the late $3d$ TM (Co-Cu) oxides, $10Dq$ wins and the TM ion takes the low spin (LS) configuration. The important point is that Co^{3+} is just located at the LS side of the border of the HS/LS ground state, and therefore, $R^{3+}CoO_3$ (R = rare earths) barely takes the LS state in the ground state. This fact implies that (possibly multiple) magnetic states would be located just above the LS ground state and they would be thermally populated at higher temperatures. Moreover, the Co^{3+} ion has six $3d$ electrons that will completely fill the three t_{2g} states in the LS state, meaning that this $t_{2g\uparrow}{}^3 t_{2g\downarrow}{}^3$ LS (described as ${}^1A_{1g}$ in the group theory term) ground state is the completely nonmagnetic one that appears only once in the series of $R^{3+}MO_3$ ($M = 3d$ TM). This harmonic coincidence, realized only in Co^{3+}, is the fundamental origin of the temperature-induced spin crossover phenomena in $LaCoO_3$ and other $RCoO_3$ in terms of electronic structure [3].

Tanabe-Sugano diagram [4] can vividly visualize such situations. Figure 2.1a shows the Tanabe-Sugano diagram for the d^6 system, in which a set of the horizontal line and an upward line (labeled as ${}^5F_2(d\varepsilon^4 d\gamma^2)$) denotes the HS (${}^5T_{2g}$: $t_{2g\uparrow}{}^3 t_{2g\downarrow}{}^1 e_{g\uparrow}{}^2$) state, a set of a downward line and the horizontal line (labeled as ${}^1A_1(d\varepsilon^6)$) denotes the LS (${}^1A_{1g}$) state, and a set of a downward and an upward line (labeled as ${}^3F_1(d\varepsilon^5 d\gamma)$) denotes the intermediate spin (IS) (${}^3T_{1g}$: $t_{2g\uparrow}{}^3 t_{2g\downarrow}{}^2 e_{g\uparrow}{}^1$) state.

This figure tells us that $LaCoO_3$ must be located at a point just beyond the HS/LS switching point $Dq/B = 2$, and the first magnetic excited state from the LS ground state must be the HS state. As is shown in Fig. 2.1b, it is obvious that the electronic

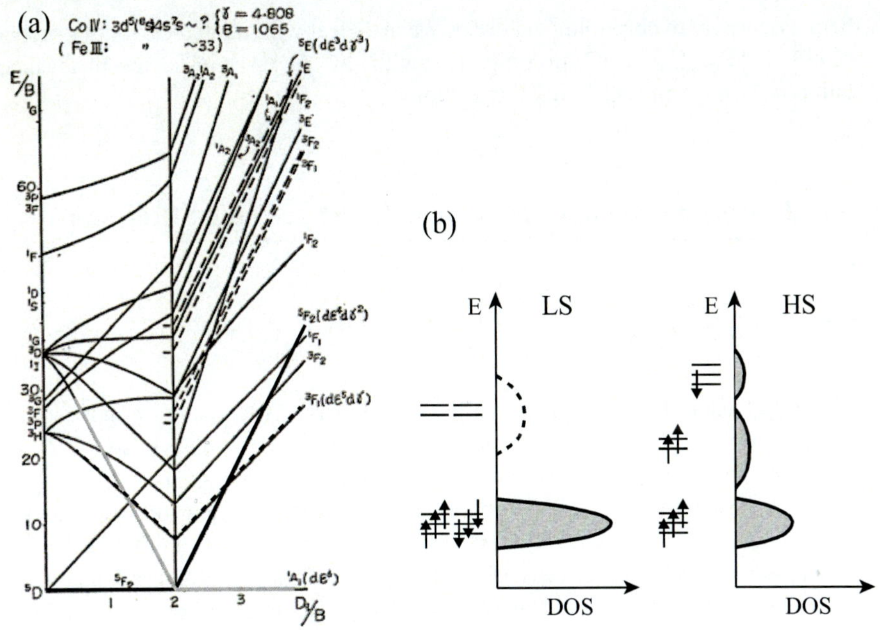

Fig. 2.1 **a** Tanabe-Sugano diagram for d^6 system [4]. The thick black line, the thick gray line, and the thick broken line (all highlighted by the author) denote the HS ($^5T_{2g}$), the LS ($^1A_{1g}$), and the IS ($^3T_{1g}$) state, respectively. Beyond $Dq/B = 2$, the ground state changes from the HS to the LS state. $LaCoO_3$ is expected to be located at a place just a little beyond this point. **b** Schematic drawing of the expected one-electron DOS in the LS and the HS state

configurations of the LS and the HS are so different that the one-electron density of states (DOS) in the valence and the conduction bands should also be very different. Such differences would be manifested in temperature-dependent changes in the valence-band PES and O K edge XAS spectra of $LaCoO_3$. In addition to them, we should also pay attention to the Co $2p$ core-level PES and Co L edge XAS because they also contain rich information on the spin configuration in the Co $3d$ states. Therefore, PES and XAS research should be very powerful to address the spin crossover problem in $LaCoO_3$. Nevertheless, history tells us that this problem is not so easy; from many magnetic susceptibility data, researchers got to know that they may have to consider another magnetic state, the IS state, as the first excited state. However, this is completely incompatible with the expected location of $LaCoO_3$ in Fig. 2.1a, because a much larger Dq/B (about 2.7) is required to realize the IS state as the first excited state. Moreover, the energy splittings between the LS, IS, and HS in Fig. 2.1a are $E_{IS} - E_{LS} > E_{HS} - E_{IS}$, which is not in agreement with the expected one from magneitc susceptibility measurements.

In this chapter, we first review the long history of electron spectroscopic studies and related theoretical studies on the spin crossover phenomena in $LaCoO_3$. To highlight the several inconsistencies of them to other experimental probes, we divide this history into two parts, before and after the idea of IS emerged, and will discuss them essentially in chronological order. We also deal with related Co oxides such as $RCoO_3$ and $Pr_{0.5}Ca_{0.5}CoO_3$ in comparison with $LaCoO_3$. We present recent advances with concluding remarks in the last sections.

2.2 Historical Review of Electron Spectroscopic Studies on $LaCoO_3$

2.2.1 *Early History*

We can go back to 1970s to find the earliest research on the temperature-induced spin crossover in $LaCoO_3$ in terms of the electronic structure. In these early years, the 100-K peak in the magnetic susceptibility had almost been ignored because this peak was often obscured by a large Curie term due to magnetic impurities. By contrast, the plateau around 500 K was always distinguished and was identified by Heikes, Miller, and Mazelsky [5] already in 1964 (Fig. 2.2). It is, therefore, reasonable that electron spectroscopic studies started with such a high-temperature range, for which the technique is, unfortunately, not suitable even nowadays.

Thornton et al. [6] measured the valence-band x-ray photoemission spectroscopy (XPS) spectra using Al Kα (1486.6 eV) x-ray source at the temperatures from 78 K to 1210 K in order to observe possible changes from the LS state to the HS state across 500 K, the temperature of the plateau in magnetic susceptibility measurements. Following the scenario by Bhide et al. [7], they interpreted the huge changes they observed as considerable charge transfer between the LS and HS Co^{3+} ions giving rise to LS Co^{2+} and HS Co^{4+} ions. From the modern point of view, however, it is

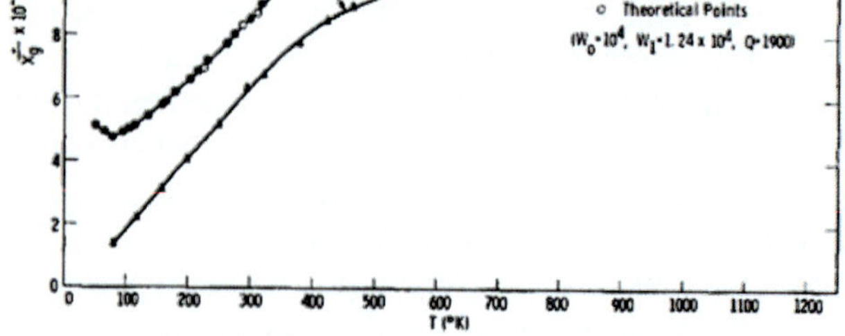

Fig. 2.2 Reciprocal magnetic susceptibility of $La_{1-x}Sr_xCoO_3$ (x=0 and 0.05) [5]

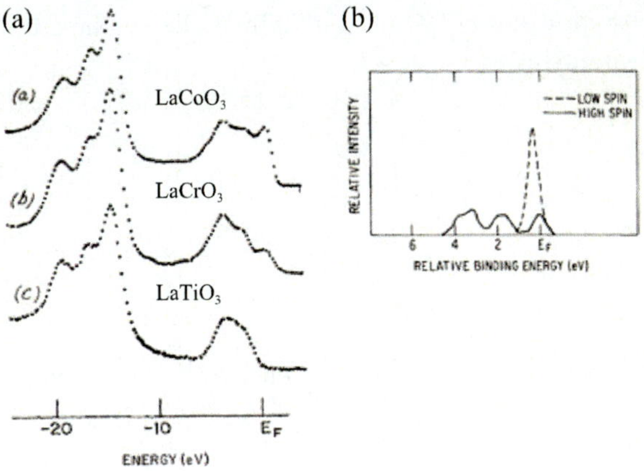

Fig. 2.3 **a** Valence-band XPS spectra of $LaCoO_3$ compared with $LaCrO_3$ and $LaTiO_3$ [8]. **b** Final-state multiplet spectra of the LS and the HS states. This figure is a reproduction of the original figure by Veal and Lam [8] by Richter et al. [9]. Note that the spectra are essentially similar to the schemtic DOS in Fig. 2.1b

difficult to reconcile this interpretation with the accumulation of investigations by many other techniques so far, and the observed changes would be attributed to oxygen deficiency at the sample surface produced by the combination of high temperatures and the ultrahigh vacuum.

Veal and Lam [8] measured XPS spectra of $LaCoO_3$ at room temperature and 573 K and analyzed the spectra using an ionic multiplet model. This was the first multiplet analysis on photoemission spectra of $LaCoO_3$. As is obvious in Fig. 2.3, the multiplet theory predicted very different valence-band electronic structures depending upon whether the system is in the LS or HS states and they concluded that the LS and the HS states coexisted at room temperature.

Overcoming the difficulties in high-temperature measurements with careful surface preparations, Richter, Bader, and Brodsky attempted to measure photoemission spectra in a very wide temperature range again from 140 to 1050 K [9], they succeeded in observing temperature-dependent intensity changes of the 1-eV peak in the valence-band ultraviolet photoemission spectroscopy (UPS) spectra, which was due to the Co $3d$ t_{2g} states in LS. The intensity of the 1-eV peak decreased with increasing temperature, which was a signature of decrease in the LS state. However, they observed no intensity changes elsewhere. Considering the high surface sensitivity of UPS, they proposed a surface reconstruction model in which the ground state of the surface Co atom could be in the HS state.

2.2.2 *Configuration-Interaction Cluster-Model Analysis*

In 1990s, a new multiplet analysis that included the oxygen $2p$ states had been developed. This is a configuration-interaction (CI) cluster model [10], in which both the Co $3d$ and the O $2p$ states in the CoO_6 octahedral cluster are taken into account. The model Hamiltonian is described by the three electronic-structure parameters, the charge-transfer energy Δ, the $3d-3d$ Coulomb interaction U, and the O $2p$–Co $3d$

hybridization $(pd\sigma)$. Within this scheme, the ground (=initial) state of $LaCoO_3$ can be written as

$$|g\rangle = a_1|d^6\rangle + a_2|d^7\underline{L}\rangle + a_3|d^8\underline{L}^2\rangle + \cdots \tag{2.1}$$

and final states of the valence-band and core-level PES can be written as

$$|f_{\mathrm{VB}}\rangle = \alpha_1|d^5\rangle + \alpha_2|d^6\underline{L}\rangle + \alpha_3|d^7\underline{L}^2\rangle \cdots \tag{2.2}$$

$$|f_{\mathrm{core}}\rangle = \beta_1|\underline{c}d^6\rangle + \beta_2|\underline{c}d^7\underline{L}\rangle + \beta_3|\underline{c}d^8\underline{L}^2\rangle \cdots . \tag{2.3}$$

Then the photoemission intensity I is described as

$$I_{\mathrm{VB}} = |\langle f_{\mathrm{VB}}|g\rangle|^2 = |\alpha_1 a_1 + \alpha_2 a_2 + \alpha_3 a_3 + \cdots|^2 \tag{2.4}$$

$$I_{\mathrm{core}} = |\langle f_{\mathrm{core}}|g\rangle|^2 = |\beta_1 a_1 + \beta_2 a_2 + \beta_3 a_3 + \cdots|^2 . \tag{2.5}$$

By changing the value of Δ, U, and $(pd\sigma)$ (meaning that the coefficients a, α, and β vary) in order to obtain the best fit to the experiment, one can deduce the reasonable values of Δ, U, and $(pd\sigma)$.

Chainani, Mathew and Sarma conducted multiple spectroscopic research using XPS, UPS, and bremsstrahlung isochromat spectroscopy (BIS) of $La_{1-x}Sr_xCoO_3$ [11]. They estimated the band gap of $LaCoO_3$ to be about 0.6 eV from the UPS and BIS spectra. They also performed the first CoO_6 CI cluster-model calculation on the Co $2p$ XPS spectrum (Fig. 2.4), from which they obtained the values of Δ, U_{dd}, and $(pd\sigma)$ as 4.0, 3.4, and 2.2 eV, respectively [12].

In 1993, Abbate et al. reported the first temperature-dependent Co $L_{2,3}$ edge and O K edge XAS studies on $LaCoO_3$ in combination with a valence-band XPS study from 80 to 630 K [13]. They found that both higher and lower energy tails of the Co L_3 edge XAS spectra showed very characteristic changes with temperature; Fig. 2.5a shows that the feature A, which is distinct at low temperatures (300 and 80 K) loses its intensity at high temperatures (550 and 630 K) while the feature B oppositely develops at high temperatures. Figure 2.5b shows that the atomic multiplet calculations assuming the LS and the HS ground state can essentially explain the 300 K and the 630 K spectra. It should be noted that the important role of the Co L_3 edge XAS spectrum in this problem has been established by this study and checking the features A and B is now a standard analysis to judge the spin state of Co oxides. Being consistent with the Co L_3 edge XAS spectra, the O K edge XAS spectra showed the same temperature dependence; the spectrum was a single peak at 300 and 80 K but a new structure appeared in the lower photon energy side of the peak at 550 and 630 K (Fig. 2.5c). This change should simply be interpreted as that the original single peak and the new structure correspond to the unoccupied O $2p$–Co $3d$ hybridized e_g and t_{2g} states, respectively. (See also Fig. 2.1b.) In accordance with this change, the valence-band XPS spectrum in Fig. 2.5d showed almost no change from 80 K to 300 K, and only a small spectral weight loss of the 1-eV peak at 570 K. Therefore, they interpreted these changes as a temperature-induced LS-to-HS transition at the 500 K plateau. Abbate et al. also compared their 80-K valence-band spectrum with

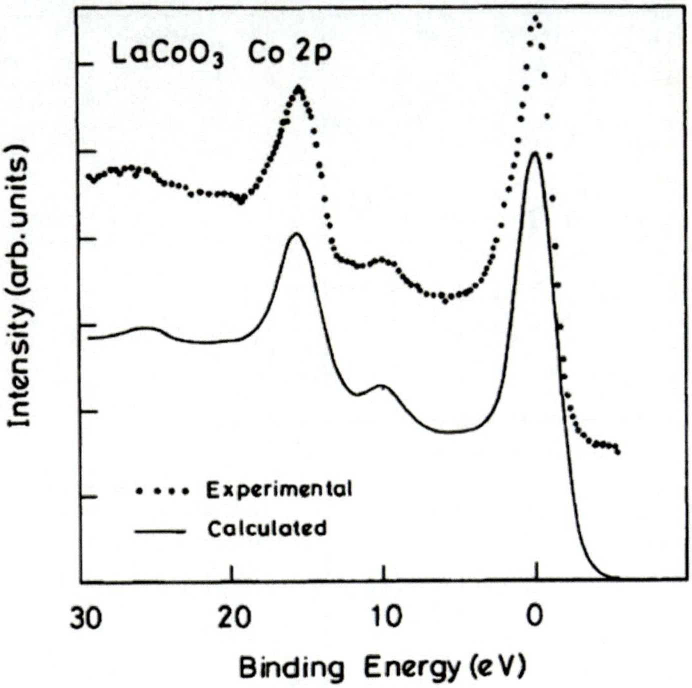

Fig. 2.4 Co $2p$ XPS spectrum of $LaCoO_3$ compared with a CI cluster-model calculation [11]. Note that the horizontal axis shows the binding energy with respect to the $2p_{3/2}$ main peak

the band-structure calculations and the CoO_6 CI cluster-model calculations. From the comparison with the CI cluster-model calculations, they obtained 4.0 eV, 5.0 eV, and −1.5 eV for Δ, U, and $(pd\sigma)$, respectively [14]. On the other hand, however, it was being recognized by the early 1990s that the LS-to-HS transition occurs around 100 K, as confirmed by neutron diffraction and NMR studies [15–17]. Hence, their report had aroused much controversy.

2.3 A New Scenario-Intermediate Spin State

Until 1990s, the fact that there are two Curie terms in χ-T plot, meaning two different magnetic states, had not been recognized well, and each study had mostly regarded one of the two temperature regions as the HS state. However, a few studies had already noticed this issue; Heikes, Miller, and Mazelsky obtained two magnetic moments from their data (Fig. 2.2), the smaller one at the lower temperature region and the larger one at the higher temperature region. To explain this, they introduced the idea of the IS state for the first time [5]. Jonker explained his χ^{-1}-T plot from 50 to 1000 K using a LS-HS thermal population model [18] proposed by Naiman et

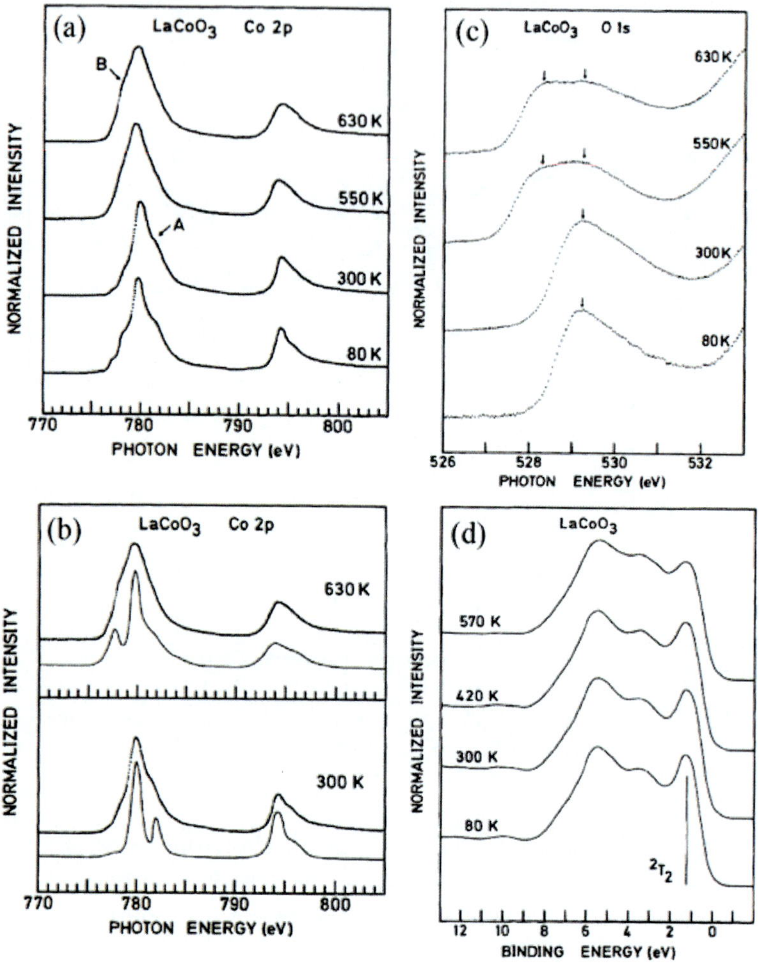

Fig. 2.5 **a** Co L edge XAS spectra of $LaCoO_3$ at various temperatures. **b** Comparisons between the Co L edge XAS spectra at 300 and 630 K (thick lines), and those of atomic multiplet calculations assuming the LS and the HS ground state (thin lines), respectively. **c** O K edge XAS spectra of $LaCoO_3$ at various temperatures. **d** XPS spectra of $LaCoO_3$ at various temperatures. All figures from [13]

al., which originally included only the LS ground state and the lowest branch of the $^5T_2$ state due to spin-orbit interaction (SOI) in order to explain the low-temperature magnetic moment [19]. Because the nonmagnetic ground state had been established by a neutron scattering measurement in the very early stage [20], it was getting widely accepted that the spin crossover problem includes two magnetic states and that the 1st excited state has a smaller magnetic moment than the 2nd one since the recognition of the 100-K anomaly in the χ-T plot as a spin crossover in 1990s.

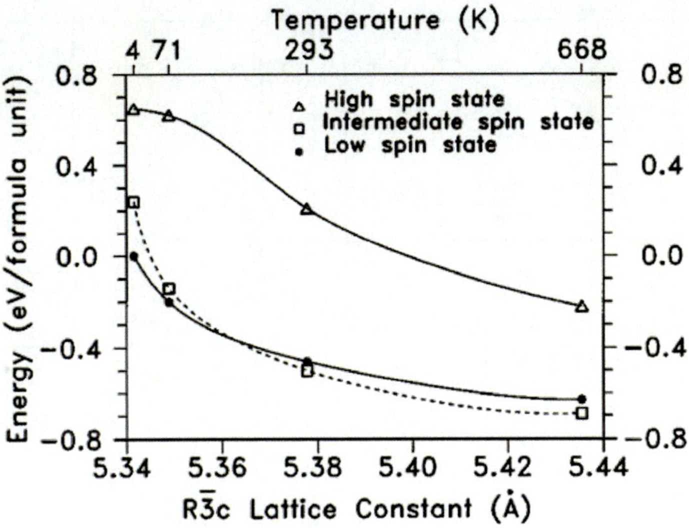

Fig. 2.6 The total energies for the HS and IS states of $LaCoO_3$ relative to the energy of the LS (t_{2g}^6) state at 4 K plotted against the $R\bar{3}c$ lattice constant. The corresponding temperature scale is shown in the upper horizontal axis [23]

Ignoring SOI, the first three electronic states are LS ($^1A_{1g}$), HS ($^5T_{2g}$), and IS ($^3T_{1g}$). To explain the susceptibility data, IS must be lower than HS, which can never occur in ordinary situations according to the Tanabe-Sugano diagram (Fig. 2.1a). However, as the Tanabe-Sugano diagrams were originally calculated for complex ions and isolated transition-metal impurities in solids [21], the energy differences between these spin states may be overestimated when the diagram is applied to solid states. In fact, while the IS energy from the ground state in the Tanabe-Sugano diagram (Fig. 2.1a) is always larger than ~1 eV [22], Korotin et al. theoretically demonstrated for the first time that the IS state was stabilized and much lower than the HS state using LDA+U band-structure calculations [23] (Fig. 2.6). They also pointed out the importance of orbital ordering in the IS state.

Mizokawa and Fujimori performed spin- and orbital-unrestricted Hartree-Fock calculations and also showed that the energy differences among the three states were able to be smaller than the Tanabe-Sugano diagram [24].

Saitoh et al. conducted comprehensive experimental and theoretical electronic structure research on $LaCoO_3$ [25]. Figure 2.7 shows the magnetic susceptibility of $LaCoO_3$ analyzed using the single-site LS-HS ($^1A_1$-$^5T_2$), LS-IS ($^1A_1$-$^3T_1$), and LS-IS-HS ($^1A_1$-$^3T_1$-$^5T_2$) models. According to their analysis, the LS-HS model cannot explain the height of the 100-K peak (Fig. 2.7a) while the LS-IS-HS model can essentially reproduce the χ-T graph in a very wide temperature range from 0 K beyond 1000 K (Fig. 2.7b). The population of each spin state deduced from this analysis is shown in Fig. 2.7c.

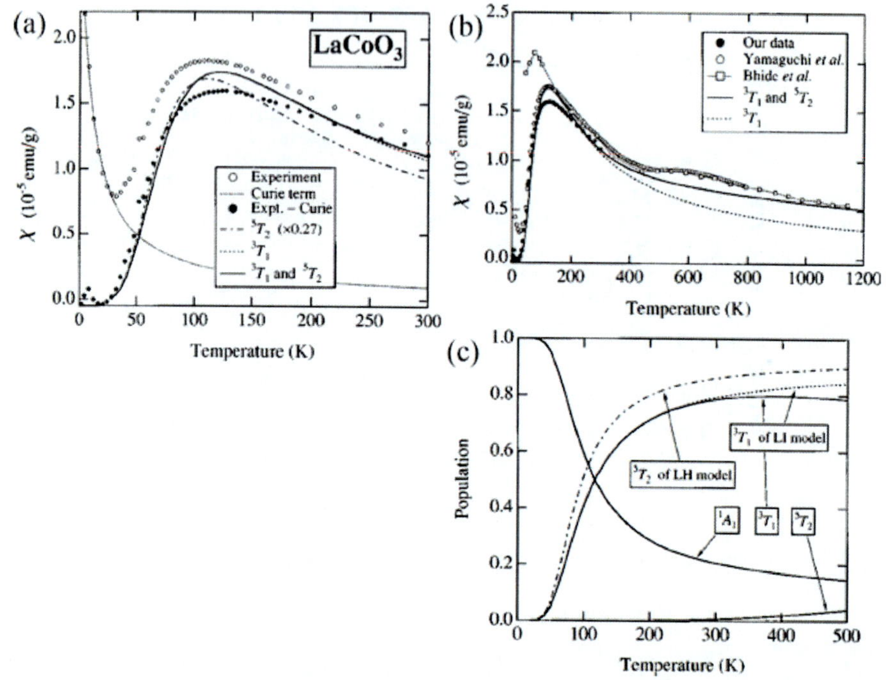

Fig. 2.7 **a** Magnetic susceptibility of $LaCoO_3$ measured at 10 kOe together with the LS-HS, LS-IS, and LS-IS-HS model analyses. **b** LS-IS and LS-IS-HS models are compared with the reported magnetic susceptibility data[7, 26] in a wider temperature range. **c** Polulation of the $^5T_2$ level in the LS-HS model (dot-dashed line), the $^3T_1$ level in the LS-IS model (dashed line), and the $^1A_1$, $^3T_1$, and $^5T_2$ levels in the LS-IS-HS model. All figures are from [25]

They performed the CoO_6 CI cluster-model calculations for the Co $2p$ core-level photoemission, the valence-band photoemission, and the O K edge XAS spectra so that the calculations reproduced the experiment. They obtained 2.0 eV, 5.5 eV, and -1.8 eV for the values of Δ, U, and $(pd\sigma)$, respectively, in the $^1A_1$ ground (=initial) state. These values are somewhat different from those by other authors [11, 14], but Δ is smaller than U in all the works, meaning that $LaCoO_3$ is categorized into a charge-transfer insulator. From the calculations, they concluded that the LS ground state of $LaCoO_3$ had heavily mixed d^6 and $d^7\underline{L}$ character, reflecting the strong covalency.

Using these values, they also performed the CI cluster-model calculations on the valence-band photoemission assuming the two other initial states, $^3T_1$, and $^5T_2$. The results are shown in Fig. 2.8a. One can see that the IS spectrum [(2)] is rather similar to the LS spectrum [(1)] and the HS spectrum [(3)] is very different from them. In fact, the on-off difference spectrum of the $3p-3d$ resonant photoemission at 80 K (lower panel of Fig. 2.8a) can be explained by the LS or LS-IS calculation (Fig. 2.8c) apart from the enhancement of the satellite. Taguchi et al. also measured the $3p-3d$ resonant photoemission spectra at 20 K and 110 K, and obtained very similar result

[27]. They also observed small spectral weight transfer with increasing temperature from the near-E_F to higher binding- energy range in the on-off difference spectrum, which they interpreted corresponds to the 100-K spin crossover.

Based on the population of the LS and IS state deduced from the analysis in Fig. 2.7c, Saitoh et al. calculated the valence-band spectrum in the scheme of CI cluster-model+He I (Fig. 2.8b(1)) and CI cluster-model+O $2p$ partial DOS (Fig. 2.8b(2)), both of which showed a good agreement with the experiment at 80 K. This is not very surprising because except for the satellite structure the band-structure calculations assuming the LS ground state [28] also well reproduced the experiment (Fig. 2.8c) as Abbate et al. [14]. Saitoh et al. also calculated temperature dependence of the simulation spectra assuming the LS-IS(-HS) model and the LS-HS model based on the population analysis in Fig. 2.7c. The result shown in Fig. 2.8d demonstrates that the LS-HS model would show appreciable changes from 80 to 300 K. From all the above analysis, they concluded that the LS-IS-HS model was most promising.

Not until recently have appreciable changes of the O K edge XAS spectrum in the temperature range up to 300 K been found [13, 25]. In 2001, however, Toulemonde et al. observed obvious changes in the O K edge XAS spectrum from 20 to 300 K in 2001 [29]. This is probably because all the measurements before them had been done in the total electron yield (TEY) mode, whereas Toulemonde et al. employed the fluorescence yield mode, which is more bulk sensitive than TEY. Figure 2.9a shows their results. Comparing the spectra with the schematic diagram (Fig. 2.9b), they interpreted the feature A and B at 20 K as the $e_{g\uparrow}$ and $e_{g\downarrow}$ unoccupied states in the LS electron configuration, respectively, and the new feature C at 300 K as the $t_{2g\uparrow}$ unoccupied state in the IS electron configuration. From this analysis, they concluded that the 100-K peak in the magnetic susceptibility is due to the LS-IS spin crossover.

Magnuson et al. investigated the temperature-dependent evolution of the electronic structure of $LaCoO_3$ using resonant (soft) x-ray emission spectroscopy (RXES) and ab initio band-structure calculations within the fixed-spin moment method [30]. Comparing with the experimental data and the *ab initio* calculations, they concluded that the system showed the LS-IS spin crossover and it remained in the IS state up to 510 K. Because XES is a much more bulk-sensitive technique than PES and even XAS in the total electron yield mode, this is a strong evidence for the LS-IS-HS scenario.

Even recently, there have been reports that support the LS-IS-HS scenario. For example, Feygenson et al. inferred that the LS-to-IS crossover occurred from 150 to 550 K based on their extended x-ray absorption fine structure, x-ray, and neutron diffraction measurements [31].

2.4 The Ground State Electronic Structure of $LaCoO_3$

Whether the first excited state is in the HS or the IS or even another state, it is established that the ground state electronic structure of $LaCoO_3$ is in the LS ($^1A_1$)

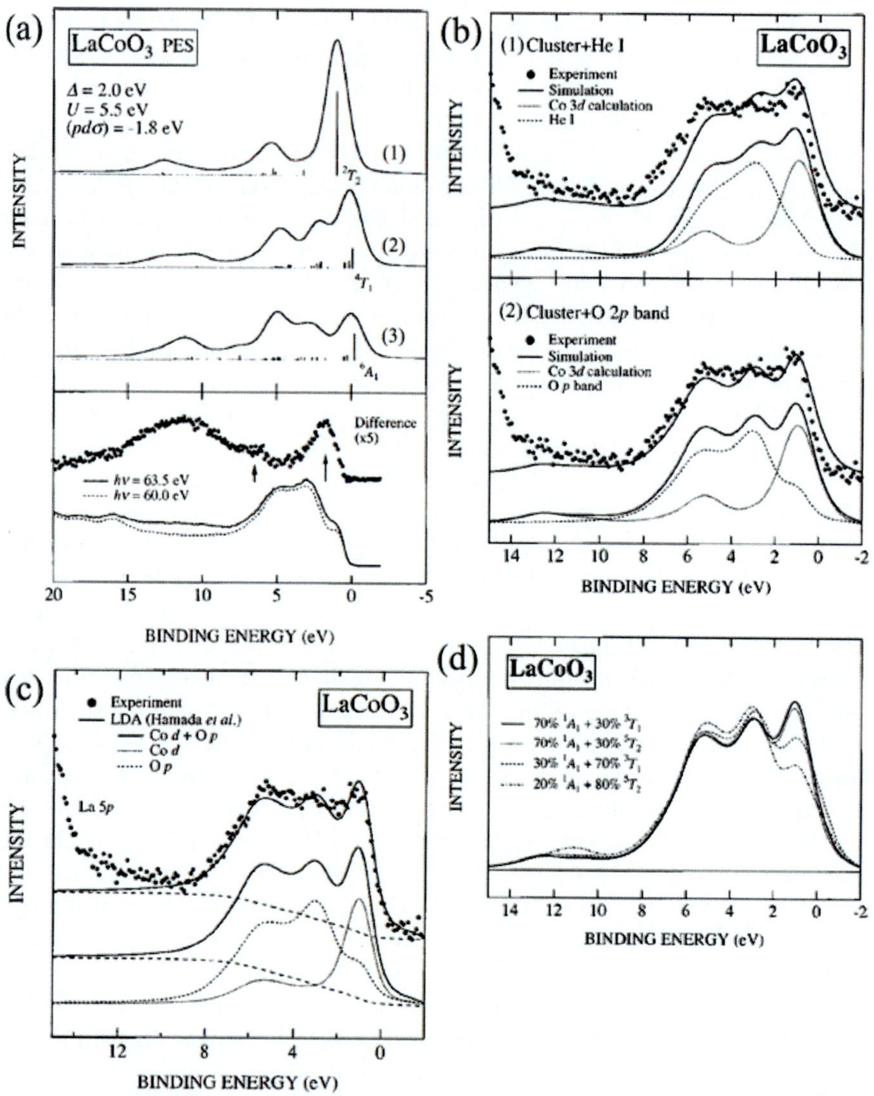

Fig. 2.8 **a** Upper panel: Full-multiplet CI cluster-model calculations of the Co 3*d* photoemission of $LaCoO_3$ assuming the (1) $^1A_1$, (2) $^3T_1$, and (3) $^5T_2$ initial state. The symmetry of the first ionization states of the spectra is also shown. Lower panel: Co $3p-3d$ resonant photoemission spectra and the on-off difference spectrum. **b** Full-multiplet CI cluster-model calculations assuming the $70\%^1A_1 + 30\%^3T_1$ initial states compared with analyses of the valence-band XPS spectra of $LaCoO_3$ taken at 80 K. (1) The O 2*p* partial DOS is simulated with a He I photoemission spectrum. (2) The O 2*p* partial DOS is from the calculations by Hamada et al. [28]. **c** Valence-band XPS spectrum of $LaCoO_3$ taken at 80 K compared with the LDA band-structure calculation by Hamada et al. [28]. **d** Simulations of the temperature-dependent XPS spectra of $LaCoO_3$ for the mixed $^1A_1$-$^3T_1$ and the mixed $^1A_1$-$^5T_2$ initial states. $70\%^1A_1 + 30\%^3T_1$ and $70\%^1A_1 + 30\%^5T_2$ correspond to $\sim$80 K and $30\%^1A_1 + 70\%^3T_1$ and $20\%^1A_1 + 80\%^5T_2$ correspond to $\sim$300 K. All figures are from [25]

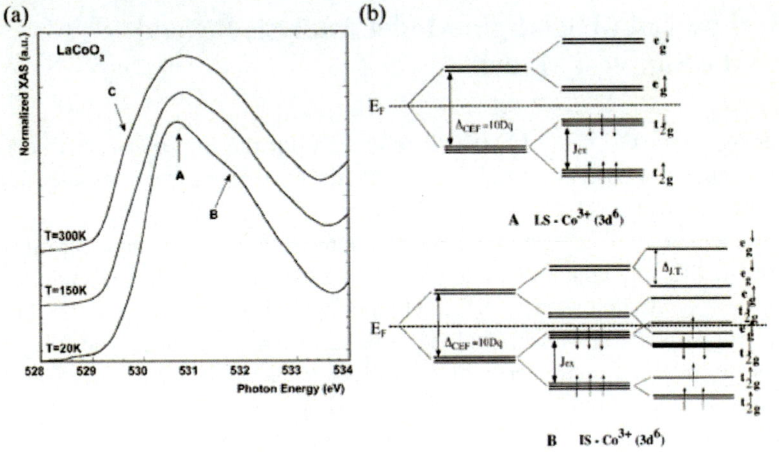

Fig. 2.9 **a** Temperature-dependent O K-edge XAS of $LaCoO_3$ measured in the fluorescence yield mode. **b** Schematic diagram of the electronic structure of $LaCoO_3$ in the LS and IS state. All figures are from [29]

state. Here, let us briefly summarize the features of the LS ground state electronic structure.

A typical valence-band photoemission spectrum of $LaCoO_3$ has four major structures, A (~0.8 eV), B (~3 eV), C (~5 eV), and D (~7 eV) [13, 14, 25, 32, 33] (See Figs. 2.5, 2.11, 2.12, and 2.14). The leading peak A, which is the most relevant to the physical properties of $LaCoO_3$, corresponds to the Co $3d$ t_{2g} band in the band picture or the Co $3d^5$ $^2T_2$ final state in the CI cluster-model. Such an intense t_{2g} peak is due to the t_{2g}^6 LS electron configuration with as many as six electrons, and hence cannot be observed in any other $3d$ transition-metal oxides.

In general, both LDA/GGA(+U) band-structure calculations (with no or moderate U) and CI cluster-model calculations can reproduce many of the valence-band features. In particular, it is common that the Co $3d$ partial density of states (the Co $3d^5$ final-state distribution in CI cluster-model calculations) has a large peak at A and a small peak at C [14, 25, 34] (see Fig. 2.8).

The CI cluster-model analyses have shown that $LaCoO_3$ is a charge-transfer insulator [11, 14, 25]. The large enhancement of the satellite in the on-off difference spectra of the $3p-3d$ resonant photoemission [25, 27] is a manifestation of the charge-transfer insulator nature. A combination of the small Δ and the intense and narrow Co $3d$ t_{2g} band gives rise to large effective hybridization and strong covalency. Hence, the ground state wave function $|g\rangle$ includes considerable weight of $|d^7\underline{L}\rangle$ and $|d^8\underline{L}^2\rangle$. Indeed, the weight of $|d^7\underline{L}\rangle$ is about 50% and the net d electron number is as large as 6.9 [25], meaning that there exist a considerable amount of O $2p$ holes.

2.5 Low Spin-High Spin Model Analysis Beyond the Simplest Model

After the proposal of the LS-IS-HS model and/or the importance of the IS state, several analyses based on the LS-HS scenario beyond the simple ionic model had been proposed.

Noguchi et al. investigated the excited states of $LaCoO_3$ using electron spin resonance (ESR) and concluded that the first excited state has the effective spin of 1, which was consistent with a HS ($^5T_2$) branch with the fictitious spin of 1 due to SOI [35]. Ropka and Radwanski analyzed this ESR data and succeeded in explaining the origin of the effective spin of 1 (namely triplet spin) state [36]. These works demonstrated the importance of SOI although the estimated magnetic susceptibility by Noguchi et al. was still three times higher than the observed one [35].

From the viewpoint of the electronic structure, the effects of SOI on the whole valence-band spectrum should be much smaller than the differences between the LS-HS crossover and the LS-IS crossover because the energy scale of SOI is no more than a few tens of meV, and this fact has so far justified the valence-band analyses using the LS, IS, and HS spectra and their superpositions without considering SOI. Therefore, although SOI would be able to explain the magnetic susceptibility, it may not be enough to explain the temperature-dependent changes in the photoemission and x-ray absorption spectra.

Kyômen et al. proposed a negative cooperative effect in order to explain the small magnetic susceptibility [37, 38]. This model is based on the idea by Naiman et al. [19] and Goodenough and coworkers [39–41]. In this model, the excitation energy from the LS ground state to the HS state(s) increases with the increasing temperature, and hence increasing number of HS Co sites due to a repulsive inter-HS site interaction. Employing this model and the energy diagram by Ropka and Radwanski [36], they successfully explained the experimental magnetic susceptibility and the heat capacity.

Haverkort et al. investigated the temperature-dependent electronic structure of $LaCoO_3$ using XAS and magnetic circular dichroism (MCD) at the Co $L_{2,3}$ edge [42]. Figure 2.10(a) compares the experimental Co $L_{2,3}$-edge XAS spectra of $LaCoO_3$ at between 20 and 650 K compared with the corresponding theoretical spectra calculated using a CoO_6 CI cluster-model in the LS-HS scenario.

2.6 Bulk Versus Surface Contribution to Electron Spectroscopic Data of $LaCoO_3$

Although PES and XAS are very powerful experimental techniques to investigate the electronic structure of solids, it should also, however, be noted that PES (and sometimes XAS as well) is in general surface sensitive so that the surface contribution to the spectra may obscure the bulk contribution to the spectra. This issue has recently

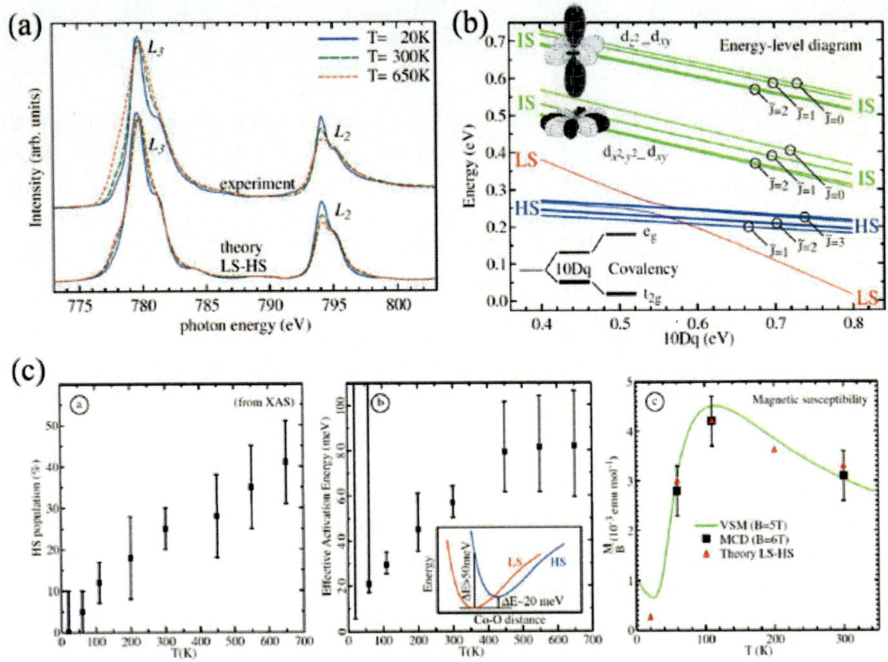

Fig. 2.10 **a** Experimental Co $L_{2,3}$-edge XAS spectra of $LaCoO_3$ at between 20 and 650 K (only three temperatures are shown) compared with the corresponding theoretical spectra calculated using a CoO_6 CI cluster-model in the LS-HS scenario. **b** Energy level diagram of a CoO_6 cluster as a function of the $10Dq$ added to the crystal field splitting due to hybridization. **c** (c-a) HS population from XAS data. (c-b) Effective activation energy from the LS to the lowest HS state. (c-c) Measured magnetic susceptibility was compared with those from the cluster-model calculations (triangles) using the HS population in Panel (c-a) and from MCD data (squares). All figures are from Ref. [42]

turned out to be important to understand the electron spectroscopic data of $LaCoO_3$ because there have been increasing evidences that the surface of bulk polycrystals [43, 44], single crystals [45], and epitaxial thin films [46] of $LaCoO_3$ is ferromagnetic below $T_C \cong 85$ K.

We have two strategies to eliminate the surface contribution to the spectra as below.

2.6.1 Bulk/surface Contribution Analysis on PES Spectra Using Different Emission Angles

The photoemission intensity at a given electron kinetic energy and an emission angle θ (with respect to the normal) can be described as

$$I(\theta) = \int_0^{\infty} \rho(z)\, e^{-\frac{z}{\lambda \cos\theta}} dz\,, \tag{2.6}$$

where z is the depth of the sample, λ is the electron escape depth, and $\rho(z)$ is the photoelectron intensity with this kinetic energy yielded at depth z. If the electronic structure of the sample varies with z, $\rho(z)$ would vary with z. Since largest changes should appear at and near the surface, one can approximate $\rho(z)$ to be two constant parts, namely the surface layer contribution ρ_s and the bulk contribution ρ_b. Also approximating the surface layer thickness to be the lattice constant a, $I(\theta)$ can be a simple form

$$I(E_K, \theta) = \lambda \cos\theta \left(\rho_s \left(1 - e^{-\frac{a}{\lambda \cos\theta}}\right) + \rho_b \left(e^{-\frac{a}{\lambda \cos\theta}}\right)\right). \tag{2.7}$$

Using normal and grazing (60° off normal) emission spectra at $h\nu = 700$ eV, and based on the above analysis, Koethe has succeeded in separating the bulk and the surface contribution to the experimental spectra taken at three temperatures and has clearly shown that the large spectral changes are mostly due to bulk [32]. Figure 2.11 shows valence-band PES spectra of $LaCoO_3$ taken under several experimental conditions. Panel (a) of Fig. 2.11 shows a bulk/surface contribution analysis. Upper Panel of Fig. 2.11a shows the difference spectra of normal emission and grazing emission with a factor of 0.6 (deduced from the above analysis), which corresponds to the bulk contribution, while Lower Panel of Fig. 2.11a shows the difference spectra of grazing emission and normal emission with a factor of 0.6, which corresponds to the surface contribution. It is obvious that the 0.8-eV peak characteristic of the LS spectrum in the bulk-representative spectra grows up with decreasing temperature, which is a clear manifestation of the increasing LS state, while no significant changes with temperature were observed in the surface-representative spectra. Because of the lack of the sharp 0.8-eV peak in the surface-representative spectra, Koethe inferred that the surface of $LaCoO_3$ was in the HS state.

Figure 2.11b shows the valence-band PES spectra of a single crystal taken at 1486.6 eV (Al Kα x-ray source). One can observe the same temperature-dependent changes in the spectra from a cleaved surface (Upper Panel) as observed in the Upper Panel of Fig. 2.11a, while no changes can be seen in the spectra from a scraped surface (Lower Panel). Because the scraped surface has a larger surface area, the result of Fig. 2.11b agrees with and hence strongly supports the result of Fig. 2.11a.

The result of Fig. 2.11 is indeed consistent with the surface ferromagnetism if we assume that the surface is always in the HS (or even IS) state regardless of the temperature. In connection with the surface ferromagnetism, however, it is also noted that the true composition of Koethe's samples is $La_{0.992}Sr_{0.002}CoO_3$ in order to avoid possible charging effects at low temperatures. The 0.2% doping of Sr is so low that the magnetic susceptibility curve is still very similar to that of $LaCoO_3$ [32]. However, it was also reported that even such a tiny Sr-doping creates magnetic polarons with very high spin numbers [26]. Although the Koethe's experiments and analysis look convincing, this might be a possible concern because the lightly Sr-doped samples can

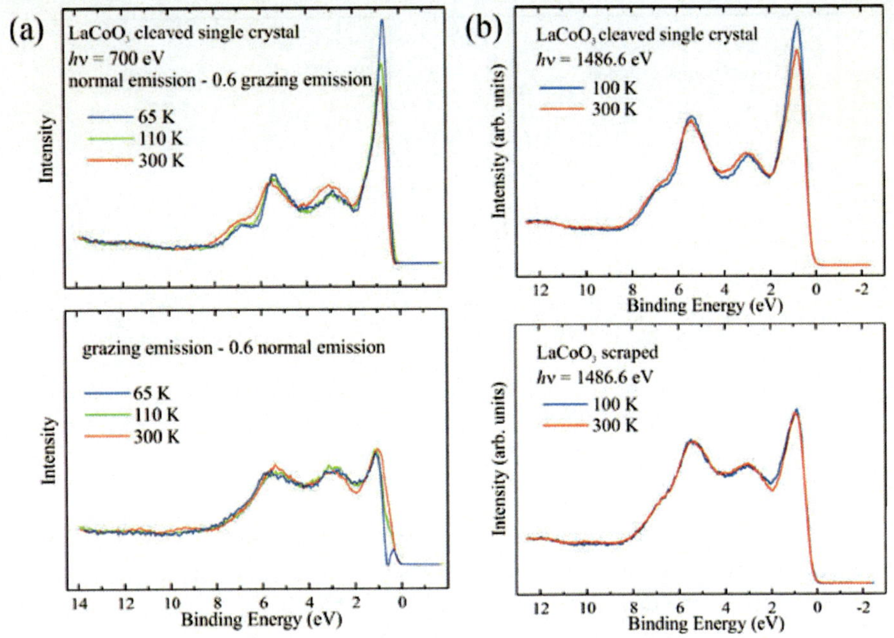

Fig. 2.11 Valence-band PES spectra of $LaCoO_3$ taken under several experimental conditions together with bulk/surface contribution analysis [32]. **a** Valence-band PES spectra taken at 700 eV after bulk/surface contribution analysis. Upper panel: Difference spectra of normal emission and grazing emission with a factor of 0.6, corresponding to the bulk contribution. Lower panel: Difference spectra of grazing emission and normal emission with a factor of 0.6, corresponding to the surface contribution. **b** Valence-band PES spectra of a single crystal taken at 1486.6 eV (Al Kα x-ray source). Upper panel: Spectra from a cleaved surface. Lower panel: Spectra from a scraped surface. Note that the true composition is $La_{0.992}Sr_{0.002}CoO_3$ to avoid charging effects

thus easily turn to be ferromagnetic, and along the surface ferromagnetism argument, such ferromagnetic Co ions may be accumulated more at the sample surface.

2.6.2 Bulk-Sensitive PES Spectra Using Hard X-ray Photoemission Spectroscopy

Another way to probe the bulk electronic structure using PES is to increase the incoming photon energy. The recent development of hard x-ray photoemission spectroscopy (HAXPES or HX-PES) using no less than 5 keV photon energy enables us to obtain very bulk-sensitive PES spectra owing to much longer λ at a several-keV range.

Figure 2.12 compares temperature-dependent valence-band PES spectra taken with (a) 700 eV and (b) 5931 eV photon energy [32]. Both are from cleaved surface of a single crystal $LaCoO_3$ at normal emission. In Fig. 2.12a, one can observe a rather

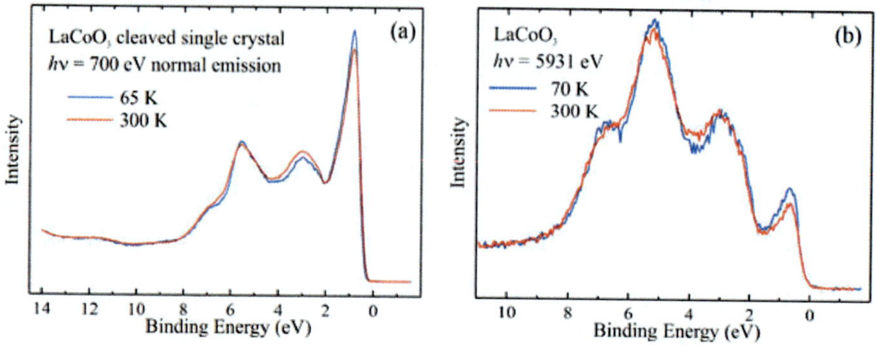

Fig. 2.12 Valence-band PES spectra of cleaved surface of a single crystal $LaCoO_3$ taken at normal emission with **a** 700 eV photon energy and **b** 5931 eV photon energy [32]. Note that the true composition is $La_{0.992}Sr_{0.002}CoO_3$ to avoid charging effects

small increase in the intensity of the 0.8-eV peak with decreasing temperature, which can be interpreted as a sum of the changing bulk contribution and the non-negligible temperature-independent surface contribution as shown in Fig. 2.11a. Compared with Fig. 2.12a, the intensity change up to 2 eV in HAXPES spectra in Fig. 2.12b is largely enhanced. This should be attributed to the much higher bulk sensitivity due to much longer λ (more than 50 Å). Thus, not only for $LaCoO_3$, HAXPES is getting more and more popular to investigate the bulk electronic structure of solids.

2.7 Recent Advances in the Electron Spectroscopic Studies

2.7.1 *Polarization-Dependent Hard X-Ray Photoemission Spectroscopy*

The angular distribution of photoelectrons is depend upon the (linear) polarization vector (namely, the electric field vector) of the incident photon beam and the orbital character (namely, the orbital angular momentum) of the initial state wave function. This effect is much pronounced in a hard x-ray range over ~5 keV as shown in theoretically [47–49] and experimentally [50, 51]; the photoionization cross section of s and np ($n > 4$) states are strongly suppressed in the so-called perpendicular or "vertical" geometry compared with the so-called parallel or "horizontal" one, whereas that of d, f, and np ($n \leq 4$) shows relatively weak changes. Here the parallel geometry is the most typical geometry in HAXPES measurements, in which the polarization vector and the direction of the detected photoelectrons are parallel, while they are perpendicular to each other in the perpendicular geometry, which can be realized by changing the direction to the detector [51, 52] or the polarization vector [50, 53]. This effect is in fact a polarization vector-emission angle dependence of the photoionization cross section. Nevertheless, it is usually called the polarization dependence.

Very recently, Takegami et al. have discovered a significant contribution of the semi-core La $5p$ intensity to the valence-band HAXPES spectrum of $La_{0.99}Sr_{0.01}CoO_3$ and have identified the origin as a combination of the very large La $5p$ photoionization cross section in the hard x-ray range and its enhancement due to the parallel geometry [52]. This method may enable us to investigate the role of La $5p$ orbitals in the physical properties of $LaCoO_3$.

2.7.2 DMFT-based IAM Calculations for Co $2p$ Core-Level Spectra

Recent advances of HAXPES has also renewed the conventional understanding of the TM $2p$ core-level spectra; Horiba et al. have discovered a new feature in the lower binding energy side of the $2p_{3/2}$ main peak in the Mn $2p$ HAXPES spectra of $La_{1-x}Sr_xMnO_3$ [54]. Employing an extended CI cluster model, they interpreted this new feature as a well-screened peak which is characteristic of bulk electronic structure, and hence had not been observed in the conventional XPS measurements. After this discovery, a series of similar rich structures in TM $2p$ core-level HAXPES spectra have been reported [55, 56].

In the above model, an electronic level that represented the metallic state at E_F was phenomenologically introduced. Alternatively, Hariki et al. have employed the impurity Anderson model (IAM) based on the dinamical mean-field therory (DMFT) to explain these rich structures [57–59] by focusing on the nonlocal screening effects [60]. Figure 2.13 shows their DMFT-based IAM and the CoO_6 CI cluster-model calculations. The Co $2p_{3/2}$ peak in the HS spectra is found to be much broader than that in the LS spectra, particularly in the DMFT-based IAM calculations. This method has thus provided a new precise analysis on the spin crossover phenomena of $LaCoO_3$.

2.7.3 Resonant Inelastic X-Ray Scattering

Resonant inelastic x-ray scattering (RIXS, which has essentially the same meaning as RXES,) has become more and more powerful owing to a rapid improvement of the energy resolution. Recently, Tomiyasu et al. have carried out RIXS measurements on $LaCoO_3$ in the temperature range of 20–550 K [61], whose energy resolution has been much improved compared with the experiment by Magnuson et al. [30]. Combining the experiment with cluster-model calculations, they concluded that the spin crossover was from LS to HS, which was associated with a change of the effective Coulomb energy strength, the ratio of the Coulomb energy to the bandwidth.

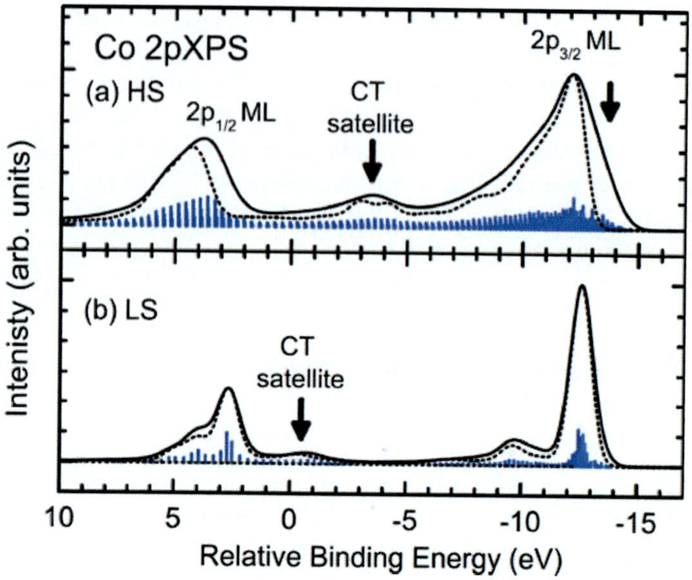

Fig. 2.13 Theoretical Co $2p$ photoemission spectra for (a) HS and (b) LS initial state calculated using the DMFT-based IAM (solid curves) and the CoO_6 CI cluster model (dotted curves). The theory is broadened using a 0.5-eV HWHM Gaussian to simulate experimental broadenings [58]

Because RIXS is bulk sensitive and hence free from the surface issues, it will become a more important tool to investigate the electronic structure of Co oxides and other compounds.

2.8 RCoO$_3$ and Other Related Co Oxides

2.8.1 PrCoO$_3$ and NdCoO$_3$

There have been few studies on the experimental elctronic structure of $PrCoO_3$ and $NdCoO_3$, compared with $LaCoO_3$.

Saitoh et al. investigated the electronic structure of $PrCoO_3$ and $NdCoO_3$ using Pr/Nd $4d-4f$ resonant photoemission spectroscopy at 200 K [33] Fig. 2.14 shows their results. The width of the leading 0.8-eV peak, which corresponds to the Co $3d$ t_{2g} states of $LaCoO_3$, is significantly larger than that of $PrCoO_3$ and $NdCoO_3$ (Fig. 2.14b). They interpreted this as a manifestation of the mixed LS and IS state in $LaCoO_3$ at 200 K because $PrCoO_3$ and $NdCoO_3$ are in the purely LS state at this temperature [62]. Figure 2.14e shows the $4d-4f$ resonant on-off difference spectra of $PrCoO_3$ and $NdCoO_3$, which experimentally represent the Pr and Nd $4f$ partial DOS. In particular, one can see a double-peak structure at about -1.6 eV in the

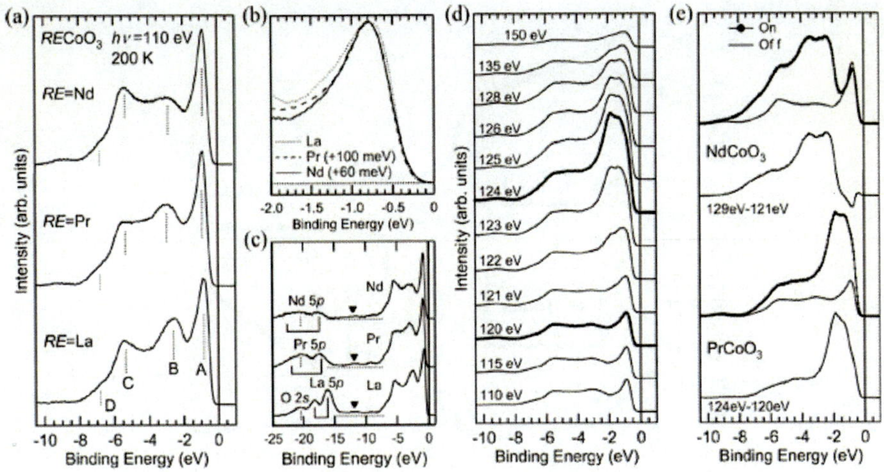

Fig. 2.14 **a** Photoemission spectra of $RECoO_3$ (RE=La, Pr, and Nd) in the valence-band region taken with 110 eV photons at 200 K. **b** Comparison of the peak width of A. **c** Spectra including shallow core levels. **d** Pr $4d-4f$ Resonant photoemission spectra of $PrCoO_3$ taken at 200 K. The on- and off-resonant spectra are highlighted by thick lines. **e** On-off difference spectra of $PrCoO_3$ and $NdCoO_3$. Note that a small dip at 0.8 eV in the Nd $4f$ difference spectrum is an artifact due to a rather large difference of the on- and off-resonant energies. All figures are from [33]

Pr $4f$ partial DOS. This makes a marked contrast to a single sharp peak at -2 eV in that of $Pr_{1-x}Sr_xMnO_3$ [63] and is rather similar to the leading peak in the Pr $4f$ partial DOS of $PrBa_2Cu_3O_{7-\delta}$, which has finite O $2p$–Pr $4f$ hybridization [64, 65]. From this analogy, Saitoh et al. pointed out importance of the O $2p$–Pr (Nd) $4f$ hybridization in Pr(Nd)CoO_3. In fact, the O $2p$–Pr (Nd) $4f$ hybridization is enhanced by isovalent substitution of Y for Pr in $PrCoO_3$; Kanai et al. have recently observed evolution of the peak intensity ratio in the double-peak structure in Pr $4d-4f$ and $3d-4f$resonant photoemission spectra of $Pr_{1-x}Y_xCoO_3$ [66], which supports the ligand hole transfer model of this system [67].

Pandey and coworkers investigated the electronic structure of $PrCoO_3$ using Mg/Al $K\alpha$ XPS, UPS, and LDA+U or GGA+U band-structure calculations. [68–70] They also reported that the Pr $4f$ states were strongly hybridized with the O $2p$ and Co $3d$ states. However, they concluded that the GGA+U calculations assuming the IS state gave a good representation of the experimental data at 300 K [70]. They also investigated the unoccupied electronic state of $LaCoO_3$ and $PrCoO_3$ by using inverse photoemission spectroscopy and GGA+U calculations, finding a fair agreement between experiment and theory [69].

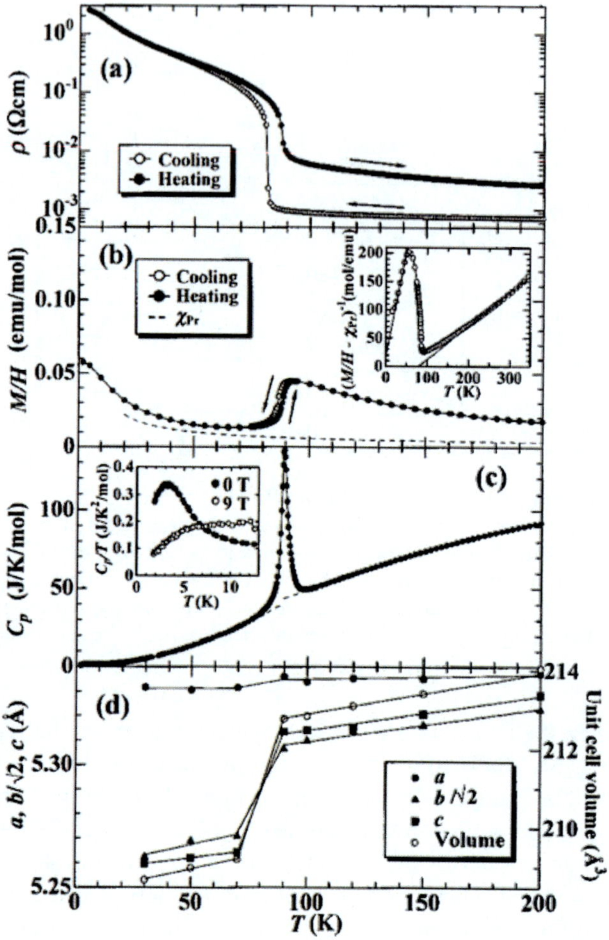

Fig. 2.15 **a** Resistivities, **b** dc magnetizations, **c** heat capacities, and **d** lattice constants and unit cell volume of $Pr_{0.5}Ca_{0.5}CoO_3$ [71]

2.8.2 $Pr_{0.5}Ca_{0.5}CoO_3$ and Related Co Oxides

Tsubouchi et al. discovered very curious phase transitions in $Pr_{0.5}Ca_{0.5}CoO_3$ [71]. Figure 2.15 shows temperature dependence of various physical properties of $Pr_{0.5}Ca_{0.5}CoO_3$. An anomaly at 90 K is evident in all the panels, demonstrating a first-order simultaneous metal-insulator and paramagnetic-paramagnetic spin-state transition. From these measurements, they interpreted this transition as the one between the localized $t_{2g}^6 + t_{2g}^5$ (Co^{3+}+Co^{4+}) in the low-temperature phase and the itinerant $t_{2g}^5 e_g^{0.5}$ in the high-temperature phase.

Saitoh et al. investigated how the electronic structure changed over this simultaneous transition [72]. Figure 2.16 shows temperature-dependent valence-band photoe-

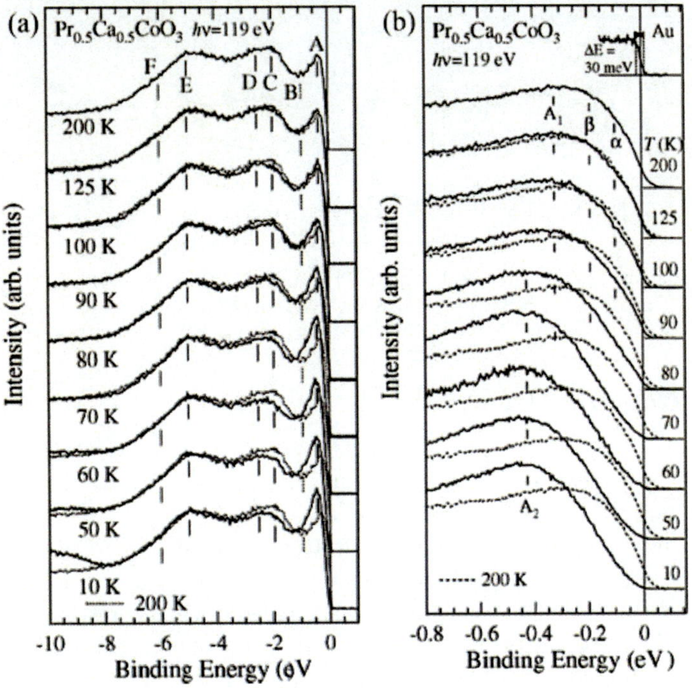

Fig. 2.16 Temperature-dependent valence-band photoemission spectra of $Pr_{0.5}Ca_{0.5}CoO_3$ taken at $h\nu = 119$ eV (a) in the whole valence-band range and (b) in the near-E_F range [72]

mission spectra of $Pr_{0.5}Ca_{0.5}CoO_3$. The photon energy was set to the off resonance of the Pr $4d-4f$ resonance to minimize the Pr $4f$ contribution (, see also Fig. 2.14). One can observe that with increasing temperature, the sharp Co $3d$ peak at $\sim$0.4 eV rapidly loses its spectral weight above $\sim$80 K and this weight is transferred to the near-E_F region to form small coherent spectral weight. They interpreted this transition as a LS-to-IS spin-state transition accompanied by an insulator-to-metal transition.

The simultaneous metal-insulator and spin-state transition in $Pr_{0.5}Ca_{0.5}CoO_3$ was so subtle that many groups have failed to reproduce the results. However, it was found that the same type of transitions emerged by applying pressure [73] and/or doping another R with a smaller ionic radius [74]. $(Pr_{1-y}Y_y)_{0.7}Ca_{0.3}CoO_3$ is a latter example. Figure 2.17 shows the Pr L_3 edge XAS spectra of $(Pr_{1-y}Y_y)_{0.7}Ca_{0.3}CoO_3$ [75]. The results show a significant valence change of the Pr ions in these samples, from 3.0+ to 3.15+ and 3.27+ for $y = 0.075$ and 0.15, respectively, which is a vivid manifestation of the Pr $4f$-O $2p$-Co $3d$ hybridization.

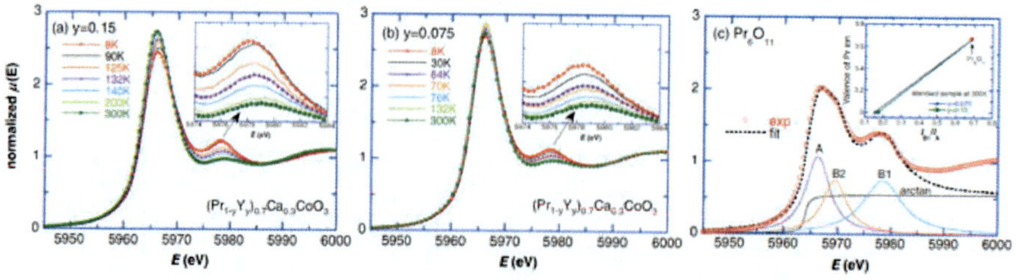

Fig. 2.17 (a) and (b) Temperature-dependent Pr L_3 edge XAS spectra of $(Pr_{1-y}Y_y)_{0.7}Ca_{0.3}CoO_3$ for $y = 0.15$ and $y = 0.075$. (c) Pr L_3 edge XAS spectra of Pr_6O_{11} at 300 K as a reference [75]

2.8.3 Other Co Oxides

There are still more literature that should be reviewed in this article. The carrier (hole)-doing systems such as $La_{1-x}Sr(Ca)_xCoO_3$ [11, 76–79] and its end $SrCoO_3$ [80] are important to understand the electronic structure of $RCoO_3$. Layered Co oxides Na_xCoO_2 [81–83] and Li_xCoO_2 [84, 85], which show superconductivity/high thermoelectricity and is a cathode of rechargeable battery, respectively, are also important because the Co ions in these compounds are basically in the Co^{3+} LS electronic configuration. There have been a few reports on the electronic structure of HS Co oxides with CoO_5 pyramidal coordination [86, 87]. Although the local environment at the Co site is different, they are also relevant to the theme of this article.

2.9 Concluding Remarks and Future Perspective

We have reviewed the experimental electronic structure of $LaCoO_3$ and related Co oxides in basically chronological order and from the viewpoint of the conflict between the LS-HS and the LS-IS-HS models. Because every experimental probe has its own advantages/disadvantages, the long-standing issue has not been settled yet. The above review article is, however, highlighting some important issues for future investigations. First is SOI. It may not significantly affect the whole valence-band electronic structure, but does affect the susceptibility and metal-insulator transition. Hence, a unified understanding of the small (=near-E_F) and the large (=whole valence band) energy-scale issues would become important. Second is beyond the single-site model. Many experiments have shown that the conventional single-site spin excitation model does not work. Hence, a good microscopic multiple-site model that has a reasonable electronic-structure origin is expected. Third is relevance to the excitonic insulator model [88]. In fact, this model may also be relevant to the above-mentioned multiple-site mode because the excitonic model also considers an adjacent site to construct an excitonic state.

Acknowledgements This article has emerged from the collaborations and innumerable discussions with my collaborators and colleagues. In particular, I would like to thank Prof. Fujimori, Prof. Mizokawa, Prof. Abbate, Prof. Kobayashi, Prof. Asai, Prof. Kyômen, and Prof. Tomiyasu, for their collaboration and the fruitful discussions.

References

1. A. Damascelli, Z. Hussain, Z.X. Shen, Rev. Mod. Phys. **75**, 473 (2003)
2. You can detect the spin polarization of an electron state in a ferromagnetic compound using spin- (and angle-) resolved PES (SR-(AR)PES), which is not the case for $LaCoO_3$
3. Historically, the temperature-induced paramagnetism observed in $RCoO_3$ has often been referred as "spin-state transition" although the phenomenon is not a phase transition but a crossover. We use these two terms rather interchangeably in order to understand the history as well as the physics
4. Y. Tanabe, S. Sugano, J. Phys. Soc. Jpn. **9**, 766 (1954)
5. R.R. Heikes, R.C. Miller, R. Mazelsky, Physica **30**, 1600 (1964)
6. G. Thornton, A.F. Orchard, C.N.R. Rao, J. Phys. C **9**, 1991 (1976)
7. V.G. Bhide, D.S. Rajoria, G. Rama Rao, C.N.R. Rao, Phys. Rev. B **6**, 1021 (1972)
8. B.W. Veal, D.J. Lam, J. Appl. Phys. **49**, 1461 (1978)
9. L. Richter, S.D. Bader, M.B. Brodsky, Phys. Rev. B **22**, 3059 (1980)
10. A. Fujimori, F. Minami, Phys. Rev. B **30**, 957 (1984)
11. A. Chainani, M. Mathew, D.D. Sarma, Phys. Rev. B **46**, 9976 (1992)
12. U_{dd} was actually deduce from an analysis of Co $L_3M_{45}M_{45}$ Auger spectrum. Hence this value should be compared with U_{eff} in other works [13,14,25]. Because $U_{\text{eff}} - U \sim -1.5$ eV for d^6 trivalent ion [89], Δ would be smaller than U as in other works
13. M. Abbate, J.C. Fuggle, A. Fujimori, L.H. Tjeng, C.T. Chen, R. Potze, G.A. Sawatzky, H. Eisaki, S. Uchida, Phys. Rev. B **47**, 16124 (1993)
14. M. Abbate, R. Potze, G.A. Sawatzky, A. Fujimori, Phys. Rev. B **49**, 7210 (1994)
15. K. Asai, P. Gehring, H. Chou, G. Shirane, Phys. Rev. B **40**, 10982 (1989)
16. K. Asai, O. Yokokura, N. Nishimori, H. Chou, J.M. Tranquada, G. Shirane, S. Higuchi, Y. Okajima, K. Kohn, Phys. Rev. B **50**, 3025 (1994)
17. M. Itoh, I. Natori, S. Kubota, K. Motoya, J. Phys. Soc. Jpn **53**, 1486 (1994)
18. G.H. Jonker, J. Appl. Phys. **37**, 1424 (1966)
19. C.S. Naiman, R. Gilmore, B. DiBartolo, A. Linz, R. Santoro, J. Appl. Phys. **36**, 1044 (1965)
20. W.C. Koehler, E.O. Wollan, J. Phys. Chem. Solids **2**, 100 (1957)
21. Y. Tanabe, S. Sugano, J. Phys. Soc. Jpn.**9**, 753 (1954)
22. The unit of the vertical axis in Fig. 1(a)is the Racah's parameter $B = 1065\ \text{cm}^{-1} \cong 0.13$ eV
23. M.A. Korotin, S.Y. Ezhov, I.V. Solovyev, V.I. Anisimov, D.I. Khomskii, G.A. Sawatzky, Phys. Rev. B **54**, 5309 (1996)
24. T. Mizokawa, A. Fujimori, Phys. Rev. B **54**, 5368 (1996)
25. T. Saitoh, T. Mizokawa, A. Fujimori, M. Abbate, Y. Takeda, M. Takano, Phys. Rev. B **55**, 4257 (1997)
26. S. Yamaguchi, Y. Okimoto, H. Taniguchi, Y. Tokura, Phys. Rev. B **53**, R2926 (1996)
27. Y. Taguchi, K. Ichikawa, T. Katsumi, K. Jouda, Y. Ohta, K. Soda, S. Kawamata, K. Okuda, O. Aita, J. Phys.: Cond. Mater. **9**, 6761 (1997)
28. N. Hamada, H. Sawada, and K. Terakura, in *Spectroscopy of Mott Insulators and Correlated Metals*, edited by A. Fujimori and Y. Tokura (Springer-Verlag, Berlin, 1995) pp.95–116
29. O. Toulemonde, N. N'Guyen, F. Studer, A. Traverse, J. Solid State Chem. **158**, 208 (2001)
30. M. Magnuson, S.M. Butorin, C. Såthe, J. Nordgren, P. Ravindran, Europhys. Lett. **68**, 289 (2004)

31. M. Feygenson, D. Novoselov, S. Pascarelli, R. Chernikov, O. Zaharko, F. Porcher, D. Karpinsky, A. Nikitin, D. Prabhakaran, A. Sazonov, V. Sikolenko, Phys. Rev. B **100**, 054306 (2019)
32. T.C. Koethe, Bulk sensitive photoelectron spectroscopy of strongly correlated transition metal oxides. Ph.D. thesis, University of Köln (2007)
33. T. Saitoh, D. Ishii, A. Hachimura, M. Hirose, T.S. Naing, Y. Kobayashi, K. Asai, M. Nakatake, M. Higashiguchi, K. Shimada, H. Namatame, M. Taniguchi, J. Magn. Magn. Mater. **310**, 981 (2007)
34. K. Knížek, P. Novák, Z. Jirák, Phys. Rev. B **71**, 054420 (2005)
35. S. Noguchi, S. Kawamata, K. Okuda, H. Nojiri, M. Motokawa, Phys. Rev. B **66**, 094404 (2002)
36. Z. Ropka, R.J. Radwanski, Phys. Rev. B **67**, 172401 (2003)
37. T. Kyômen, Y. Asaka, M. Itoh, Phys. Rev. B **67**, 144424 (2003)
38. T. Kyômen, Y. Asaka, M. Itoh, Phys. Rev. B **71**, 024418 (2005)
39. J.B. Goodenough, J. Phys. Chem. Solids **6**, 287 (1958)
40. P.M. Raccah, J.B. Goodenough, Phys. Rev. **115**, 932 (1967)
41. M.A. Señarís Rodríguez, J.B. Goodenough, J. Solid State Chem. **116**, 224 (1995)
42. M.W. Haverkort, Z. Hu, J.C. Cezar, T. Burnus, H. Hartmann, M. Reuther, C. Zobel, T. Lorenz, A. Tanaka, N.B. Brookes, H.H. Hsieh, H.J. Lin, C.T. Chen, L.H. Tjeng, Phys. Rev. Lett. **97**, 176405 (2006)
43. J. Androulakis, N. Katsarakis, J. Giapintzakis, Phys. Rev. B **64**, 174401 (2001)
44. J.Q. Yan, J.S. Zhou, J.B. Goodenough, Phys. Rev. B **70**, 014402 (2004)
45. A. Harada, T. Taniyama, Y. Takeuchi, T. Sato, T. Kyômen, M. Itoh, Phys. Rev. B **75**, 184426 (2007)
46. D. Fuchs, C. Pinta, T. Schwarz, P. Schweiss, P. Nagel, S. Schuppler, R. Schneider, M. Merz, G. Roth, H. v. Löhneysen, Phys. Rev. B **75**, 144402 (2007)
47. M.B. Trzhaskovskaya, V.I. Nefedov, V.G. Yarzhemsky, At. Data Nucl. Data Tables **77**, 97 (2001)
48. M.B. Trzhaskovskaya, V.I. Nefedov, V.G. Yarzhemsky, At. Data Nucl. Data Tables **82**, 257 (2002)
49. M.B. Trzhaskovskaya, V.K. Nikulin, V.I. Nefedov, V.G. Yarzhemsky, At. Data Nucl. Data Tables **92**, 245 (2006)
50. A. Sekiyama, J. Yamaguchi, A. Higashiya, M. Obara, H. Sugiyama, M.Y. Kimura, S. Suga, S. Imada, I.A. Nekrasov, M. Yabashi, K. Tamasaku, T. Ishikawa, New J. Phys. **12**, 043045 (2010)
51. J. Weinen, T.C. Koethe, C.F.C.F. Chang, S. Agrestini, D. Kasinathan, Y.F. Liao, H. Fujiwara, C. Schüßler-Langeheine, F. Strigari, T. Haupricht, G. Panaccione, F. Offi, G. Monaco, S. Huotari, K.D. Tsuei, L.H. Tjeng, J. Electron Spectrosc. Relat. Phenom. **198**, 6 (2015)
52. D. Takegami, L. Nicolaï, T.C. Koethe, D. Kasinathan, C.Y. Kuo, Y.F. Liao, K.D. Tsuei, G. Panaccione, F. Offi, G. Monaco, N.B. Brookes, J. Minár, L.H. Tjeng, Phys. Rev. B **99**, 165101 (2019)
53. K. Takahashi, R. Kato, M. Okawa, T. Okuda, A. Yasui, E. Ikenaga, K. Ono, N. Hamada, T. Saitoh, J. Phys. Soc. Jpn. **88**, 074701 (2019)
54. K. Horiba, M. Taguchi, A. Chainani, Y. Takata, E. Ikenaga, D. Miwa, Y. Nishino, K. Tamasaku, M. Awaji, A. Takeuchi, M. Yabashi, H. Namatame, M. Taniguchi, H. Kumigashira, M. Oshima, M. Lippmaa, M. Kawasaki, H. Koinuma, K. Kobayashi, T. Ishikawa, S. Shin, Phys. Rev. Lett. **93**, 236401 (2004)
55. M. Taguchi, A. Chainani, N. Kamakura, K. Horiba, Y. Takata, M. Yabashi, K. Tamasaku, Y. Nishino, D. Miwa, T. Ishikawa, S. Shin, E. Ikenaga, T. Yokoya, K. Kobayashi, T. Mochiku, K. Hirata, K. Motoya, Phys. Rev. B **71**, 155102 (2005)
56. M. Taguchi, A. Chainani, K. Horiba, Y. Takata, M. Yabashi, K. Tamasaku, Y. Nishino, D. Miwa, T. Ishikawa, T. Takeuchi, K. Yamamoto, M. Matsunami, S. Shin, T. Yokoya, E. Ikenaga, K. Kobayashi, T. Mochiku, K. Hirata, J. Hori, K. Ishii, F. Nakamura, T. Suzuki, Phys. Rev. Lett. **95**, 177002 (2005)
57. A. Hariki, Y. Ichinozuka, T. Uozumi, J. Phys. Soc. Jpn. **82**, 043710 (2013)
58. A. Hariki, A. Yamanaka, T. Uozumi, J. Phys. Soc. Jpn. **84**, 073706 (2015)
59. A. Hariki, A. Yamanaka, T. Uozumi, EPL **114**, 27003 (2016)

60. M.A. van Veenendaal, G.A. Sawatzky, Phys. Rev. Lett. **70**, 2459 (1993)
61. K. Tomiyasu, J. Okamoto, H.Y. Huang, Z.Y. Chen, E.P. Sinaga, W.B. Wu, Y.Y. Chu, A. Singh, R.P. Wang, F.M.F. de Groot, A. Chainani, S. Ishihara, C.T. Chen, D.J. Huang, Phys. Rev. Lett. **119**, 196402 (2017)
62. M. Itoh, M. Mori, S. Yamaguchi, Y. Tokura, Physica B **259–261**, 902 (1999)
63. J.S. Kang, C.G. Olson, J.H. Jung, S.T. Lee, T.W. Noh, B.I. Min, Phys. Rev. B **60**, 13257 (1999)
64. J.S. Kang, S. Haffner, C.G. Olson, J.H. Kim, M.B. Maple, S.K. Kwon, B.I. Min, Phys. Rev. B **66**, 052503 (2002)
65. R. Fehrenbacher, T.M. Rice, Phys. Rev. Lett. **70**, 3471 (1993)
66. D. Kanai *et al.*, unpublished
67. Y. Kobayashi, Y. Terakado, K. Asai, J. Phys. Soc. Jpn. **83**, 104704 (2014)
68. S.K. Pandey, A. Kumar, S.M. Chaudhari, A.V. Pimpale, J. Phys.: Condens. Matter **18**, 1313 (2006)
69. S.K. Pandey, A. Kumar, S. Banik, A.K. Shukla, S.R. Barman, A.V. Pimpale, Phys. Rev. B **77**, 113104 (2008)
70. S.K. Pandey, S. Patil, V.R.R. Medicherla, R.S. Singh, K. Maiti, Phys. Rev. B **77**, 115137 (2008)
71. S. Tubouchi, T. Kyômen, M. Itoh, P. Ganguly, M. Oguni, Y. Shimojo, Y. Morii, Y. Ishii, Phys. Rev. B **66**, 052418 (2002)
72. T. Saitoh, Y. Yamashita, N. Todoroki, T. Kyomen, M. Itoh, M. Higashiguchi, M. Nakatake, K. Shimada, J. Electron Spectrosc. Relat. Phenom. **144–147**, 893 (2005)
73. T. Fujita, T. Miyashita, Y. Yasui, Y. Kobayashi, M. Sato, E. Nishibori, M. Sakata, Y. Shimojo, N. Igawa, Y. Ishii, K. Kakurai, T. Adachi, Y. Ohishi, M. Takata, J. Phys. Soc. Jpn. **73**, 1987 (2004)
74. T. Fujita, S. Kawabata, M. Sato, N. Kurita, M. Hedo, Y. Uwatoko, J. Phys. Soc. Jpn. **74**, 2294 (2005)
75. H. Fujishiro, T. Naito, S. Ogawa, N. Yoshida, K. Nitta, J. Hejtmánek, K. Knížek, Z. Jirák, J. Phys. Soc. Jpn. **81**, 064709 (2012)
76. J.P. Kemp, D.J. Beal, P.A. Cox, J. Phys.: Condens. Matter **2**, 3767 (1990)
77. R.P. Vasquez, Phys. Rev. B **54**, 14938 (1996)
78. T. Saitoh, T. Mizokawa, A. Fujimori, M. Abbate, Y. Takeda, M. Takano, Phys. Rev. B **56**, 1290 (1997)
79. W.R. Flavell, A.G. Thomas, J. Hollingworth, S. Warren, S.C. Grice, P.M. Dunwoody, C.E.J. Mitchell, P.G.D. Marr, D. Teehan, S. Downes, E.A. Seddon, V.R. Dhanak, K. Asai, Y. Kobayashi, N. Yamada, Faraday Discuss. **114**, 407 (1999)
80. R.H. Potze, G.A. Sawatzky, M. Abbate, Phys. Rev. B **51**, 11501 (1995)
81. T. Takeuchi, T. Kondo, T. Takami, H. Takahashi, H. Ikuta, U. Mizutani, K. Soda, R. Funahashi, M. Shikano, M. Mikami, S. Tsuda, T. Yokoya, S. Shin, T. Muro, Phys. Rev. B **69**, 125410 (2004)
82. M.Z. Hasan, Y.D. Chuang, D. Qian, Y.W. Li, Y. Kong, A. Kuprin, R. Fedorov, A. V. nad Kimmerling, E. Rotenberg, K. Rossnagel, Z. Hussain, H. Koh, N.S. Rogado, M.L. Foo, R.J. Cava, Phys. Rev. Lett. **92**, 246402 (2004)
83. H.B. Yang, S.C. Wang, A.K.P. Sekharan, H. Matsui, S. Souma, T. Sato, T. Takahashi, T. Takeuchi, J.C. Campuzano, R. Jin, B.C. Sales, D. Mandrus, Z. Wang, H. Ding, Phys. Rev. Lett. **92**, 246403 (2004)
84. J.P. Kemp, P.A. Cox, J. Phys.: Condens. Matter **2**, 9653 (1990)
85. J. van Elp, J.L. Wieland, H. Eskes, P. Kuiper, G.A. Sawatzky, Phys. Rev. B **44**, 6090 (1991)
86. Z. Hu, H. Wu, M.W. Haverkort, H.H. Hsieh, H.J. Lin, T. Lorenz, J. Baier, A. Reichl, I. Bonn, C. Felser, A. Tanaka, C.T. Chen, L.H. Tjeng, Phys. Rev. Lett. **92**, 207402 (2004)
87. T. Sudayama, Y. Wakisaka, T. Mizokawa, H. Wadati, G.A. Sawatzky, D.G. Hawthorn, T.Z. Regier, K. Oka, M. Azuma, Y. Shimakawa, Phys. Rev. B **83**, 235105 (2011)
88. A. Sotnikov, J. Kuneš, Sci. Rep. **6**, 30510 (2016)
89. A. Fujimori, A.E. Bocquet, T. Saitoh, T. Mizokawa, J. Electron Spectrosc. Relat. Phenom. **62**, 141 (1993)

Chapter 3
Hidden Spin-States in Cobalt Oxides: Photoinduced State and Excitonic Insulating State

Sumio Ishihara

Abstract Recent theoretical researches in correlated electron systems with the spin-state degree of freedom are reviewed. Novel electronic states which appear under the competition between the low-spin and high-spin states in perovskite cobaltites are focused on. Calculated results obtained from the two-orbital Hubbard model, an effective model for the low-energy electronic structure, and the five-orbital Hubbard model are introduced. In particular, we pay our attention to the possibilities of the following three exotic states: (i) The excitonic insulating state, in which the electronic wave function is represented by the linear combination of the low-spin and high-spin states, is introduced. The ground state and finite-temperature phase diagrams and the elementary excitations are shown. We propose the possible experimental methods to identify the excitonic insulating state in cobaltates. (ii) Possibility of the electronic phase separation by hole-carrier doping is introduced. The ground state energy as a function of carrier number indicates that a homogeneous electronic state is unstable, and the electronic state is separated into the low-spin band insulator and the high-spin ferromagnetic metallic state. A microscopic mechanism of this phase separation phenomena is discussed. (iii) A bound state between the high-spin state and the hole state induced by the photoirradiation is proposed by the complement theoretical calculations. This is termed the photoinduced high-spin polaron state. Implications of this characteristic state to the optical pump-probe experiments are discussed.

Keywords Spin-state degree of freedom · Excitonic insulator · Photoinduced state

3.1 Introduction

Spin-state degree of freedom (SSDF) in the transition-metal ions has been attracted much interest in wide classes of magnetic molecules and magnetic solids [1–3]. The different local electron configurations with different magnitudes of spin are often

Sumio Ishihara is a deceased author.

S. Ishihara (✉)
Department of Physics, Tohoku University, Sendai 980-8578, Japan
e-mail: okimoto.y.aa@m.titech.ac.jp

Y. Okimoto et al. (eds.), *Spin-Crossover Cobaltite*, Springer Series in Materials Science 305, https://doi.org/10.1007/978-981-15-7929-5_3

realized in a single magnetic ion by changing temperature and external field, which are termed the SSDF. The transitions or crossovers between the multiple spin states are often observed in the iron and cobalt ions in correlated electron magnets, biomaterials, and earth inner core materials [4–7]. The spin-state transition/crossover is attributed to an energy balance between the different electron configurations, mainly controlled by the competition between the crystalline-field splitting and the ferromagnetic Hund coupling, which stabilizes the low-spin (LS) and high-spin (HS) states, respectively.

Let us consider the spin states in the cobalt oxides with perovskite crystal structure [8–10]. The chemical formula is $R\mathrm{CoO_3}$, where R represents a rare-earth ion. The formal valence of Co is 3+ and the electron number in the outermost $3d$ orbitals is approximately 6. Due to the nearly cubic-symmetric crystalline field by the surrounding O ions, the five $3d$ orbitals are split into the doubly degenerate e_g orbitals and the triply degenerate t_{2g} ones. In the strong limits of the crystalline field (Δ) and the Hund coupling (J), the electronic configurations $(e_g)^0(t_{2g})^6$ with $S = 0$ and $(e_g)^2(t_{2g})^4$ with $S = 2$ are stabilized, respectively. These are described by the multiplets in the cubic symmetry ($\mathrm{O_h}$) group as ${}^{5\mathrm{T}}_{2}$ and ${}^{1\mathrm{A}}_{1}$, respectively.

We focus on a situation in which the above two-spin states compete with each other [11–13]. It is well-known that in the analyses of the electronic structure in a single Co ion, the transition between the HS and LS states by adjusting the ratio of Δ and J is of the first order. The $S = 1$ intermediate spin (IS) state with the electron configuration of $(e_g)^1(t_{2g})^5$ (${}^3\mathrm{T}_1$) is the second-excited spin state which competes with the LS and HS states around the transition point [14]. However, the energy of the IS state is higher than those of the LS and HS states by about $10B$ where B is the Racha parameter. The first-principle electronic structure calculation suggested that the IS state has a chance to be stabilized owing to the Jahn-Teller lattice distortion coupled with the e_g orbital degree of freedom [15–19] . However, in spite of the intensive experimental examinations, clear evidence of the IS state have not been reported so far.

In addition to the IS state, several candidate states have been proposed in the region where the LS and HS states compete. One possible state is the spin-state ordered state, where the LS and HS ions are aligned periodically [20, 21]. This is termed as the LS/HS state, from now on. Another candidate state is the excitonic insulating (EI) state. The local wave function at each site in this state is represented by a linear combination of the different spin states associated with the lowering of the local crystal symmetry. The wave function is generally given by $|\psi\rangle = a|\mathrm{LS}\rangle + b|\mathrm{IS}\rangle + c|\mathrm{HS}\rangle$ with coefficients a, b, and c. The EI state was originally proposed by narrow gap semiconductors and semimetals long time ago [22–26]. When the exciton binding energy overcomes the gap energy, there is a chance that a macroscopic number of the electron-hole pairs are condensed. Now, the EI state is revisited [27, 28] and this concept is applied to the correlated electron systems with the SSDF for recent years [29–31].

It is widely recognized that the non-equilibrium states induced by intensive light irradiation often realize hidden electronic and lattice states which are not realized in thermal equilibrium states. There is a wide class of photoinduced hidden states, and the photoinduced phase transitions have been reported so far in correlated electron systems. As for the materials with SSDF, photoinduced spin-state transitions have been studied intensively and extensively in molecular-magnet systems, such as the Prussian blue analog. One well-known examples is the Prussian blue analog $Rb_{0.43}Mn[Fe(CN)_6]_{0.81}$'E3$H_2O$, in which $Mn^{2+}(S = 5/2)$-Fe^{3+} $(S = 1/2)$ is changed into $Mn^{3+}(S = 2)$-$Fe^{2+}(S = 0)$ by light irradiation [32–39]. The intensive light irradiation effect has been also studied in strongly correlated cobalt oxides with perovskite structure with SSDF. The anomalous transient optical spectra are reported and suggest the photoinduced hidden spin states [40–45].

In this article, we review the recent theoretical researches in the correlated electron systems with SSDF. In particular, we focus on the hidden states which compete with the LS and HS states around the spin-state transition points in cobalt oxides with perovskite structure and related compounds. The theoretical models for the correlated electron systems with SSDF are introduced in Sects. 3.2 and 3.3. The EI states as a hidden state near the transition point are explained in Sect. 3.4. Effects of the magnetic field and the realistic five d-orbitals are introduced in Sects. 3.5 and 3.6, respectively. The characteristic spin and magnetic states introduced in the carrier doping are explained in Sect. 3.7. The photoinduced spin states in the correlated systems are introduced in Sect. 3.8. Section 3.9 is devoted to summary and perspective.

3.2 Two-Orbital Hubbard Model

First, we introduce the minimal theoretical models in which the correlated electron systems with SSDF and their spin crossover phenomena are able to be examined. This is the two-orbital Hubbard model, where the two kinds of the orbitals termed a and b are introduced at each site. The energy difference, $\Delta(= \varepsilon_a - \varepsilon_b) > 0$, between the two-orbital energies are introduced, and the averaged electron number per site is fixed to be two. The two-orbital Hubbard model [46–50] is defined as

$$\mathscr{H}_{\rm TH} = \mathscr{H}_t + \mathscr{H}_u. \tag{3.1}$$

The first term describes the electron hoppings between the nearest-neighbor (NN) sites given by

$$\mathscr{H}_t = -\sum_{\langle ij\rangle\eta\sigma} t_\eta (c^\dagger_{i\eta\sigma} c_{j\eta\sigma} + \text{H.c.}), \tag{3.2}$$

and the second term represents the local energy levels and the electron-electron interactions given by

Fig. 3.1 A schematic illustration of the two-orbital Hubbard model

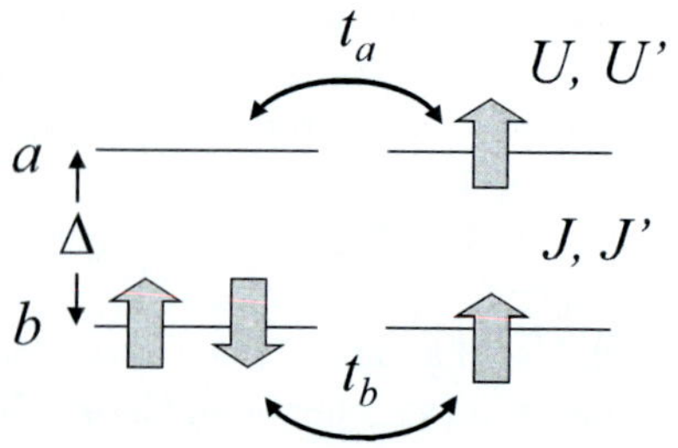

$$
\begin{aligned}
\mathscr{H}_u = \Delta \sum_i n_{ia} + U \sum_{i\eta} n_{i\eta\uparrow} n_{i\eta\downarrow} + U' \sum_i n_{ia} n_{ib} \\
+ J \sum_{i\sigma\sigma'} c^\dagger_{ia\sigma} c^\dagger_{ib\sigma'} c_{ia\sigma'} c_{ib\sigma} + I \sum_{i\eta \neq \eta'} c^\dagger_{i\eta\uparrow} c^\dagger_{i\eta\downarrow} c_{i\eta'\downarrow} c_{i\eta'\uparrow}.
\end{aligned}
\tag{3.3}
$$

A schematic picture of the energy levels and interactions in this Hamiltonian is shown in Fig. 3.1. The creation (annihilation) operator for an electron with orbital $\eta (= a, b)$ and spin $\sigma (=\uparrow, \downarrow)$ at site i is introduced as $c^\dagger_{i\eta\sigma}$ $(c_{i\eta\sigma})$. The electron number operator is defined as $n_{i\eta} = \sum_\sigma n_{i\eta\sigma}$ with $n_{i\eta\sigma} = c^\dagger_{i\eta\sigma} c_{i\eta\sigma}$. In Eq. (3.2), the electron hopping amplitudes t_η are introduced between the same orbitals in the NN sites. The first term in Eq. (3.3) represents the energy difference between the two orbitals. As for the on-site electron-electron interactions, we define the intra-orbital Coulomb interaction U, the inter-orbital Coulomb interaction U', the Hund coupling J, and the pair-hopping interaction I.

We consider the local electronic structures described by $\mathscr{H}_u$. There are the ${}_4C_2 (= 6)$ possible states at a single site with two electrons. The three kinds of the spin-singlet states are represented by

$$
|L\rangle = \left(f c^\dagger_{b\uparrow} c^\dagger_{b\downarrow} - g c^\dagger_{a\uparrow} c^\dagger_{a\downarrow} \right) |0\rangle, \tag{3.4}
$$

$$
|L'\rangle = \left(g c^\dagger_{b\uparrow} c^\dagger_{b\downarrow} + f c^\dagger_{a\uparrow} c^\dagger_{a\downarrow} \right) |0\rangle, \tag{3.5}
$$

and

$$
|L''\rangle = \left(c^\dagger_{b\uparrow} c^\dagger_{a\downarrow} - c^\dagger_{b\downarrow} c^\dagger_{a\uparrow} \right) |0\rangle, \tag{3.6}
$$

where we define $f = 1/\sqrt{1 + (\Delta' - \Delta)^2 / I^2}$, $g = \sqrt{1 - f^2}$, and $\Delta' = \sqrt{\Delta^2 + I^2}$, and the vacuum of the electron $|0\rangle$. When we set $I = 0$, we have $g = 0$ and $f = 1$. The spin-triplet states are defined as

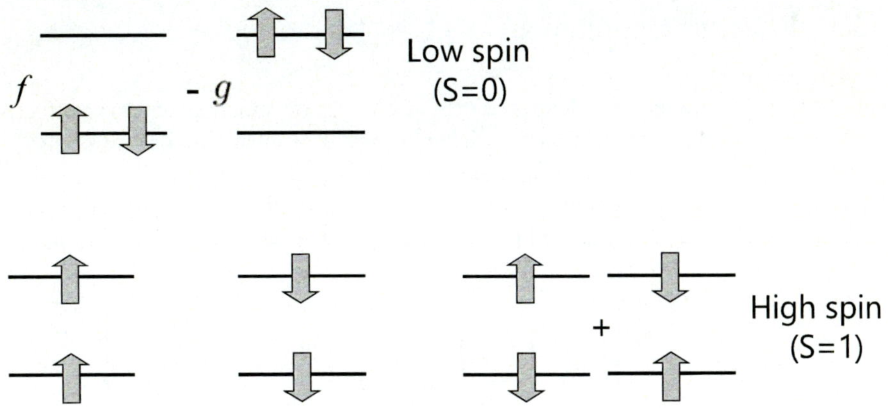

Fig. 3.2 Schematic electron configurations of the LS state (upper) and the HS state (lower)

$$|H_{+1}\rangle = c^{\dagger}_{a\uparrow} c^{\dagger}_{b\uparrow} |0\rangle, \tag{3.7}$$

$$|H_0\rangle = \frac{1}{\sqrt{2}} \left(c^{\dagger}_{a\uparrow} c^{\dagger}_{b\downarrow} + c^{\dagger}_{a\downarrow} c^{\dagger}_{b\uparrow} \right) |0\rangle, \tag{3.8}$$

$$|H_{-1}\rangle = c^{\dagger}_{a\downarrow} c^{\dagger}_{b\downarrow} |0\rangle. \tag{3.9}$$

In the case that Δ and J are comparable, $|L\rangle$ and $\{|H_{+1}\rangle, |H_0\rangle, |H_{-1}\rangle\}$ are the dominant electronic configurations in the low-energy states. It is convenient to introduce the equivalent set of the wave functions for the spin-triplet states $\{|H_X\rangle, |H_Y\rangle, |H_Z\rangle\}$, where we define

$$|H_X\rangle = \frac{1}{\sqrt{2}} \left(-|H_{+1}\rangle + |H_{-1}\rangle \right), \tag{3.10}$$

$$|H_Y\rangle = -\frac{1}{\sqrt{2}i} \left(|H_{+1}\rangle + |H_{-1}\rangle \right), \tag{3.11}$$

$$|H_Z\rangle = |H_0\rangle, \tag{3.12}$$

which will be utilized in the effective Hamiltonian introduced later. The eigen energies of these states are $E_L = U + \Delta - \Delta'$ for $|L\rangle$ and $E_H = \Delta + U' - J$ for $\{|H_{+1}\rangle, |H_0\rangle, |H_{-1}\rangle\}$. The energy difference between the two states is $E_H - E_L = \Delta - J$ in the case of $I = 0$. Schematic electron configurations of the two sets of the spin states are shown in Fig. 3.2. From now on, these configurations are termed the "LS" state ($S = 0$) and the "HS" state ($S = 1$), and the relative stability is mainly determined by the competition of Δ and J.

3.3 Effective Hamiltonian

To analyze the low-energy electronic structures, it is convenient to derive an effective model Hamiltonian obtained from the two-orbital Hubbard model introduced in Sect. 3.2. We focus on the LS and HS states as the low-energy states and restricts the Hilbert space represented by these states. We introduce the spin operators with a magnitude of $S = 1$ represented in the basis set $\{|H_X\rangle, |H_Y\rangle, |H_Z\rangle, |L\rangle\}$ defined by

$$S^x = \frac{1}{\sqrt{2}}\left(\begin{array}{ccc|c} & & & \\ & & -i & \\ & i & & \\ \hline & & & \end{array}\right), \quad S^y = \frac{1}{\sqrt{2}}\left(\begin{array}{ccc|c} & & i & \\ & & & \\ -i & & & \\ \hline & & & \end{array}\right), \quad S^z = \frac{1}{\sqrt{2}}\left(\begin{array}{ccc|c} & -i & & \\ i & & & \\ & & & \\ \hline & & & \end{array}\right). \tag{3.13}$$

We also introduce the pseudo-spin (PS) operators τ_Γ^γ with subscripts $\gamma (= x, y, z)$ and $\Gamma (= X, Y, Z)$. The X components of the PS operators, τ_X^γ, are explicitly represented as

$$\tau_X^x = \left(\begin{array}{ccc|c} & & & 1 \\ & & & \\ & & & \\ \hline 1 & & & \end{array}\right), \quad \tau_X^y = \left(\begin{array}{ccc|c} & & & -i \\ & & & \\ & & & \\ \hline i & & & \end{array}\right), \quad \tau_X^z = \left(\begin{array}{ccc|c} 1 & & & \\ & & & \\ & & & \\ \hline & & & -1 \end{array}\right). \tag{3.14}$$

The other components, τ_Y^γ and τ_Z^γ, are defined in similar ways. We note that τ_Γ^z represents a difference between the weights in the LS and HS states, and τ_Γ^x and τ_Γ^y represent the mixing between the LS and HS states with the real and complex coefficients, respectively. Thus, $\langle \tau_\Gamma^x \rangle$ and $\langle \tau_\Gamma^y \rangle$ are identified as the order parameters for the EI phase. The PS operators are also represented by the projection operators as

$$\tau_X^x = |L\rangle \langle H_X| + |H_X\rangle \langle L| , \tag{3.15}$$

$$\tau_X^y = i \left(|L\rangle \langle H_X| - |H_X\rangle \langle L| \right) , \tag{3.16}$$

$$\tau_X^z = |H_X\rangle \langle H_X| - |L\rangle \langle L| . \tag{3.17}$$

To obtain the effective model Hamiltonian, we use the standard canonical transformation where the transfer term of the Hamiltonian $\mathscr{H}_t$ is treated as the perturbed part. We derive this by following the general formula given by:

$$\mathscr{H}_{\text{eff}} = \mathscr{H}_u + \sum_\beta \frac{\mathscr{H}_t |\beta\rangle\langle\beta| \mathscr{H}_t}{E_0 - E_\beta}, \tag{3.18}$$

where $|\beta\rangle$ and E_β are the eigen state and the eigen energy for $\mathscr{H}_u$. Finally, the low-energy effective Hamiltonian for the two-orbital Hubbard model is obtained as [29, 31, 44, 45]

$$\mathscr{H}_{\mathrm{eff}} = -h_z \sum_i \tau_i^z + J_z \sum_{\langle ij \rangle} \tau_i^z \tau_j^z + J_s \sum_{\langle ij \rangle} \mathbf{S}_i \cdot \mathbf{S}_j$$
$$- J_x \sum_{\langle ij \rangle \Gamma} \tau_{i\Gamma}^x \tau_{j\Gamma}^x - J_y \sum_{\langle ij \rangle \Gamma} \tau_{i\Gamma}^y \tau_{j\Gamma}^y, \tag{3.19}$$

where we define $\tau_i^\gamma = \sum_\Gamma \tau_{i\Gamma}^\gamma$. The energy parameters are represented by the parameters in the two-orbital Hubbard model. A sign of J_s is positive, and signs of J_x and J_y reflect a sign of t_b/t_a.

In the case of $I = 0$, we have $J_x = J_y$, which implies the U(1) symmetry on the $\tau^x - \tau^y$ plane, corresponding to the relative phase degree of freedom between the wave functions of the LS and HS states. In the case of $I \neq 0$, the U(1) symmetry is reduced to the Z_2 symmetry, and the Hamiltonian is invariant under the simultaneous transformation of $\tau_i^x \to -\tau_i^x$ or $\tau_i^y \to -\tau_i^y$. The EI transition is classified into the spontaneous breaking of the Z_2 symmetry. This symmetry corresponds to the relative sign degree of freedom between the LS and HS states in the wave function.

3.4 Excitonic Insulating States and Their Dynamics

The electronic structure at zero temperature described in the effective Hamiltonian given in (3.19) was analyzed by the mean-field approximation in a square lattice [31]. The phase diagrams at $t_b/t_a = -0.1$ and -0.4 in the $\Delta - J$ planes are given in Fig. 3.3a and b, respectively. In the phase diagram shown in Fig. 3.3a, the five electronic phases are recognized. This phase diagram is obtained by calculating the HS number, $n^{\mathrm{H}} = \sum_\Gamma |H_\Gamma\rangle\langle H_\Gamma|$, the LS number $n^{\mathrm{L}} = |L\rangle\langle L|$, and the magnetic and EI order parameters, as shown in Fig. 3.4. We define the ferromagnetic and antiferromagnetic order parameters as $S_{\mathrm{F(AF)}} = (1/4) \sum_{\gamma=x,y,z} (\langle S^\gamma \rangle_A \pm \langle S^\gamma \rangle_B)^2$, where $\langle \cdots \rangle_A$ and $\langle \cdots \rangle_B$ imply the averages in the A and B sublattices, respectively,

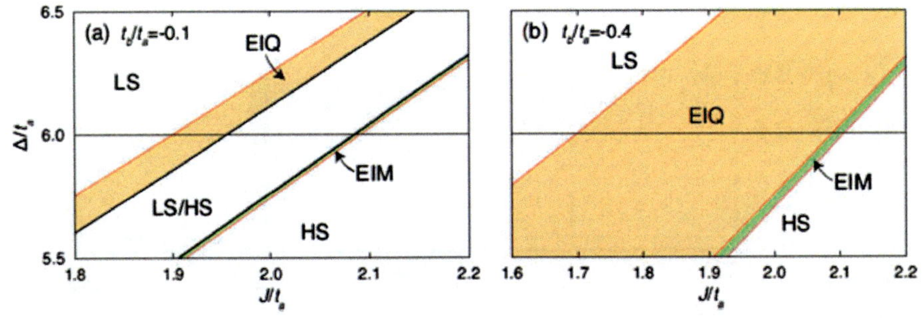

Fig. 3.3 Phase diagrams at zero temperature obtained by the effective Hamiltonian. **a** $t_b/t_a = -0.1$ and **b** -0.4. The parameter values are chosen to be $I = J$, $U = 6J$, and $U' = 4J$. Black thick lines and red thin lines represent the phase boundaries, where the phase transitions are of the first and second orders, respectively. Data were taken from [31]

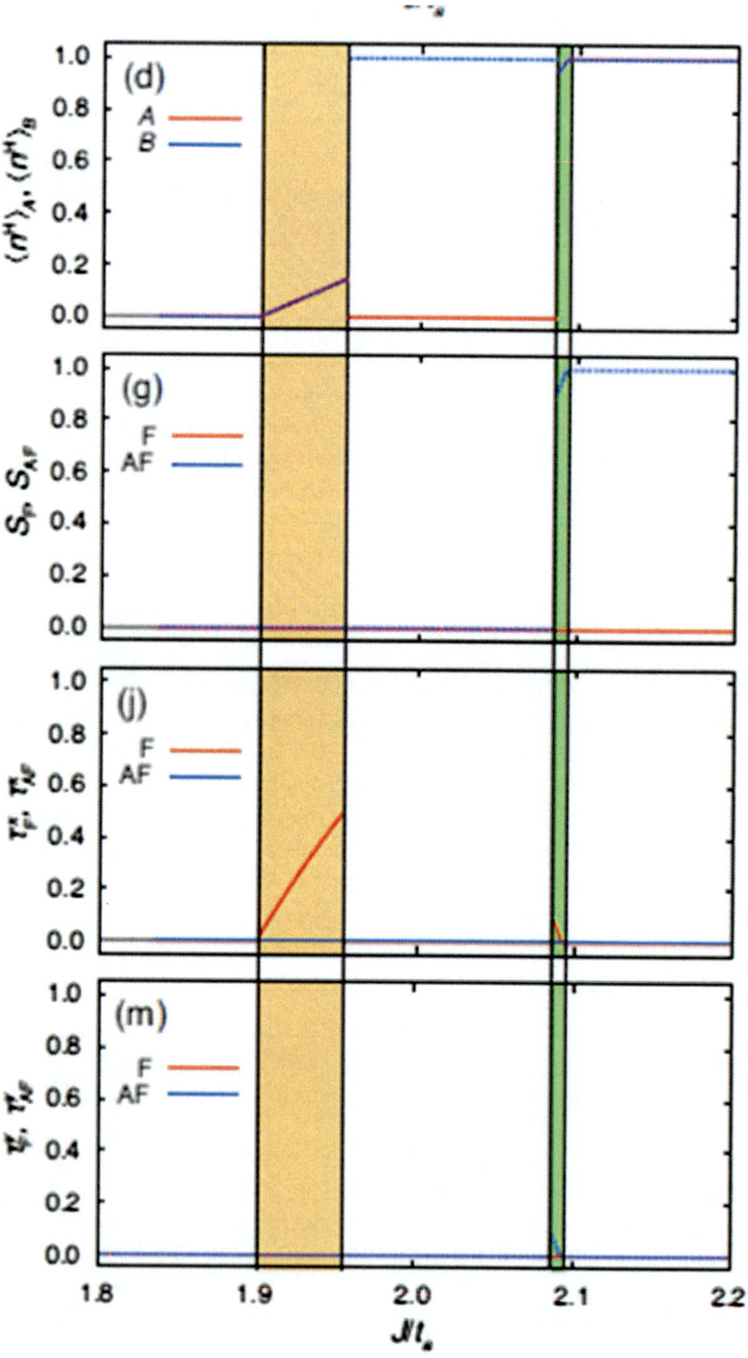

Fig. 3.4 The HS density ($\langle n^{\mathrm{H}} \rangle_A$, $\langle n^{\mathrm{H}} \rangle_B$), the spin order parameters (S_{F}, S_{AF}), the x component of the EI order parameters (τ^x_{F}, τ^x_{AF}), and the y component of the EI order parameters (τ^y_{F}, τ^y_{AF}) obtained by the effective Hamiltonian. The parameter values are chosen to be $I = J$, $U = 6J$, $U' = 4J$ and $t_b/t_a = -0.1$. Data were taken from [31]

and the EI order parameters as $\tau^{\gamma}_{\mathrm{F(AF)}} = (1/4)\sum_{\Gamma=X,Y,Z}(\langle\tau^{\gamma}_{\Gamma}\rangle_A \pm \langle\tau^{\gamma}_{\Gamma}\rangle_B)^2$. In the case of $\Delta \gg J$, $\langle n^{\mathrm{H}}\rangle = 0$, implying realization of the LS phase. On the other side, in the case of $\Delta \ll J$, $\langle n^{\mathrm{H}}\rangle = 1$, implying the HS phase, where the magnetic order parameters indicate the antiferromagnetic (AFM) order. This is termed the AFM-HS phase, from now on. Between the LS and AFM-HS phases, we find a phase which is characterized by $\langle n^{\mathrm{H}}\rangle = 0$ at the sublattice A, and $\langle n^{\mathrm{H}}\rangle = 1$ at the sublattice B. This is the LS/HS ordered state, where the HS and LS sites are aligned alternately in the square lattice. There are two EI phases located between the LS and LS/HS ordered phase, and between the AFM-HS and LS/HS ordered phase. In both the two EI phases, the LS and HS states are mixed quantum mechanically. Because of no hopping integrals between the a and b orbitals, the mixings occur spontaneously due to the electron-electron interactions. The x components, and the x and y components of the EI order parameters are finite in the EIQ and EIM phases, respectively, as shown in Fig. 3.4. Here, we focus on the EI phase between the LS/HS and LS phase, termed the EIQ phase, where $\tau^x_{\mathrm{F}} =$ is finite, implying a uniform EI phase with the real wave function. Any magnetic order parameters do not emerge. The wave function in this phase is represented by

$$|\psi_{\mathrm{EIQ}}\rangle = C_1|L\rangle + C_2|H_Z\rangle, \tag{3.20}$$

with real numbers C_1 and C_2. This EI state originates from the exchange interactions $J_x > J_y > 0$ in (3.19). In this wave function, $\langle S^x\rangle = \langle S^y\rangle = \langle S^z\rangle = 0$ and the spin-quadrupole order $\langle 3(S^z)^2 - S(S+1)\rangle \neq 0$.

Next, we introduce the electronic dynamics in the EI phase. The dynamical spin structure factors are defined as

$$S^{\gamma\gamma'}(\mathbf{q}, \omega) = \int_{-\infty}^{\infty} \frac{dt}{2\pi} \langle S_{\mathbf{q}\gamma}(t) S_{-\mathbf{q}\gamma'}\rangle e^{i\omega t}, \tag{3.21}$$

where $S_{\mathbf{q}\gamma}$ is the Fourier transform of the spin operators $S_{i\gamma}$. The corresponding static susceptibilities are defined as $\chi^{\gamma\gamma'} = \chi^{\gamma\gamma'}(\mathbf{q}, \omega = 0)\Big|_{\mathbf{q}\to 0}$, where the dynamical susceptibilities are defined by

$$\begin{aligned}\chi^{\gamma\gamma'}(\mathbf{q}, \omega) &= i\int_0^{\infty} dt \langle [S_{\mathbf{q}\gamma}(t), S_{-\mathbf{q}\gamma'}]\rangle e^{i\omega t - \epsilon t} \\ &= -\int_{-\infty}^{\infty} d\kappa \frac{S_{\gamma\gamma'}(\mathbf{q}, \kappa)}{\omega - \kappa + i\epsilon},\end{aligned} \tag{3.22}$$

with an infinitedesimal constant ε. The calculated dynamical spin correlation functions, $S^{xx}(\mathbf{q}, \omega)$ and $S^{zz}(\mathbf{q}, \omega)$, are shown in Fig. 3.5 [31]. These correspond to the transverse and longitudinal spin excitations, respectively, since the ordered moment is assumed to be parallel to the z axis. In both the EI phases, the two kinds of spin-wave modes appear. The longitudinal components are attributed to the change in the magnitude of the spin moment. This implies a change in the relative weight of the

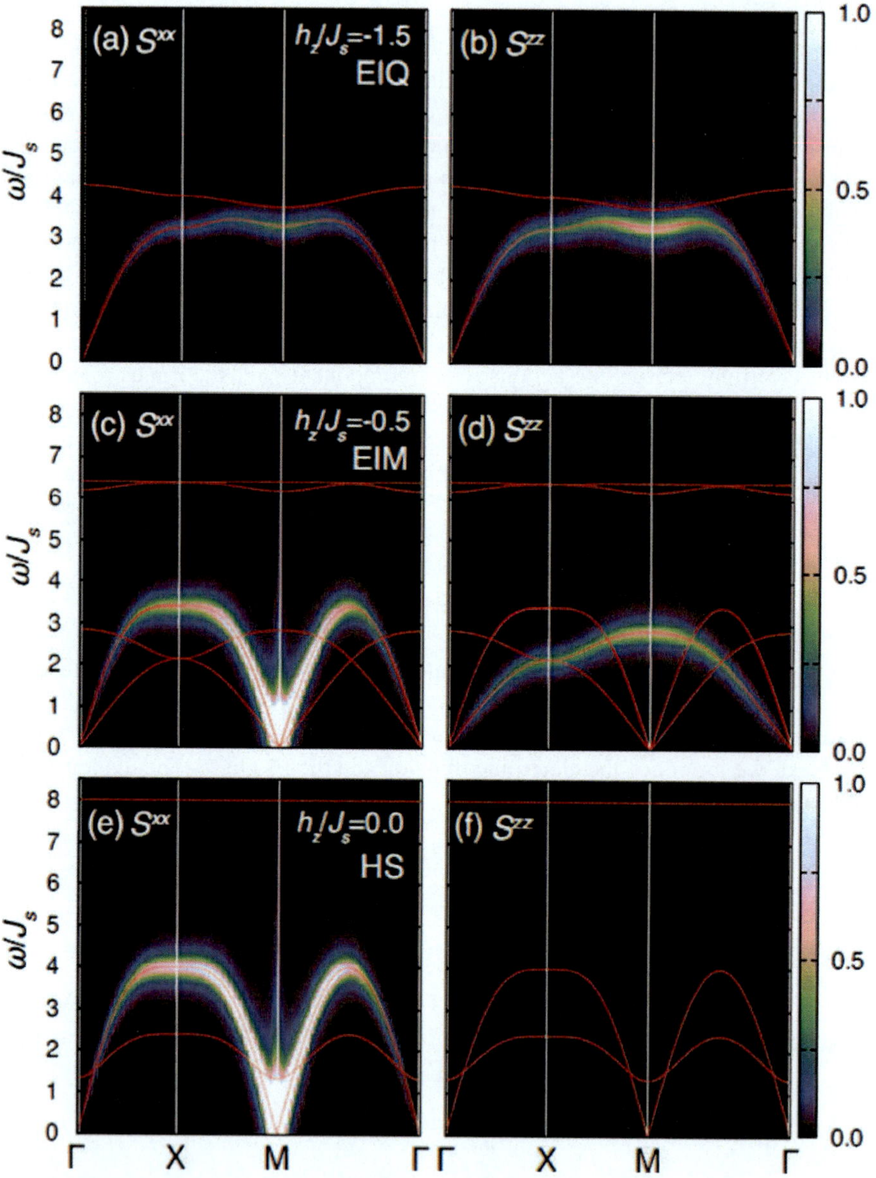

Fig. 3.5 Contour maps of the dynamical spin correlation functions calculated by the effective Hamiltonian. The x components S^{xx} and the z components S^{zz} are shown in **a**, **c**, **e** and **b**, **d**, **f**, respectively. The parameter value of h_z/J_s is taken to be **a** -1.5 (the EIQ phase), **b** -0.5 (the EIM phase), and **c** 0 (the AFM-HS phase). The dotted red lines represent the dispersion relations. Data were taken from Ref. [31]

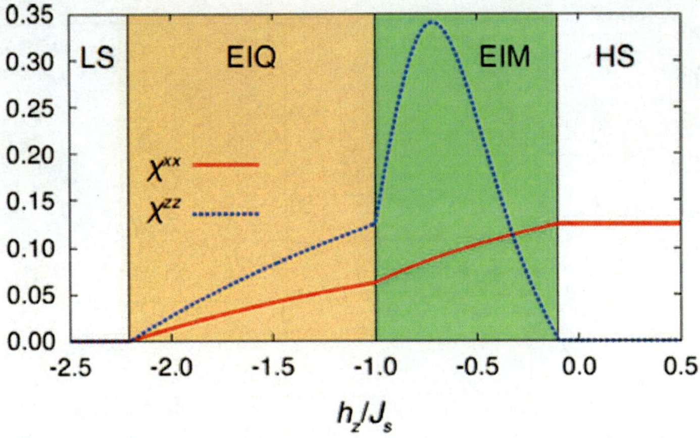

Fig. 3.6 Transverse and longitudinal components of the static magnetic susceptibilities obtained by the effective Hamiltonian. The parameter values are chosen to be $J_z/J_s = 0.1$, $J_x/J_s = 0.5$, and $J_y = 0$. Data were taken from Ref. [31]

LS and HS states in the wave function. In contrast to the results in the two EI phases, the spin excitation spectra in the HS phase have only the transverse component originating from the spin-wave excitations in the AFM order.

The transverse and longitudinal components of the susceptibilities, χ^{xx} and χ^{zz}, at zero temperature are shown in Fig. 3.6 [31]. The susceptibilities in the LS phase vanishes due to the finite spin gap. In the AFM-HS phase, $\chi^{zz} = 0$ and $\chi^{xx} \neq 0$ are attributed to the AFM order, in which the staggered moment is along the S^z axis. In the EI phases, both χ^{xx} and χ^{zz} are finite. The finite magnetic susceptibilities in the EIQ phase corresponds to the Van Vleck components. In the EIM phase, χ^{zz} is finite in contrast to the HS phase, since the longitudinal responses are attributed to a change of a relative weight of the LS and HS states.

3.5 Magnetic Field Effect

In recent years, the magnetization measurements in $LaCoO_3$ under a high magnetic field were reported [51–54]. The results are shown in Fig. 3.7b. The first-order phase transition was observed around 60T below about 30-40K. An additional first-order phase transition occurs between 80-100T at higher temperatures. Two kinds of the field-induced phase were confirmed: the low-temperature and low-field phase termed B1 and the high-field phase termed B2.

Before introducing the theoretical results for the magnetic field effect [55, 56], we show the finite-temperature phase diagram without a magnetic field presented in Fig. 3.8 [55]. The electronic structures at zero temperature is changed with increasing J as LS→EIQ→LS/HS→EIM→AFM-HS, as introduced in Fig. 3.3. With increas-

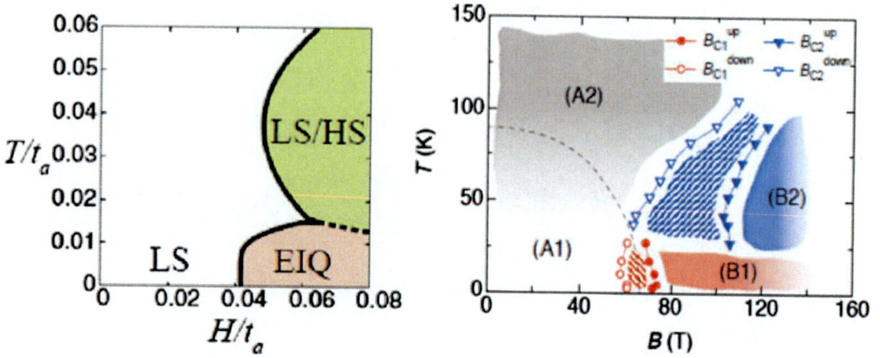

Fig. 3.7 (Left) Finite-temperature phase diagram under the magnetic field applied to the LS phase. Solid and broken lines represent the second- and first-order phase transitions, respectively. The parameter values are chosen to be $J/t_a = 4.4$, corresponding to one of the bold arrows in Fig. 3.8, and $t_b/t_a = -0.1$. Data were taken from Ref. [55]. (Right) Experimental phase diagram in $LaCoO_3$. Data were taken from Ref. [54]. Reprinted with permission from Ref. [54] ©(2016) the American Physical Society

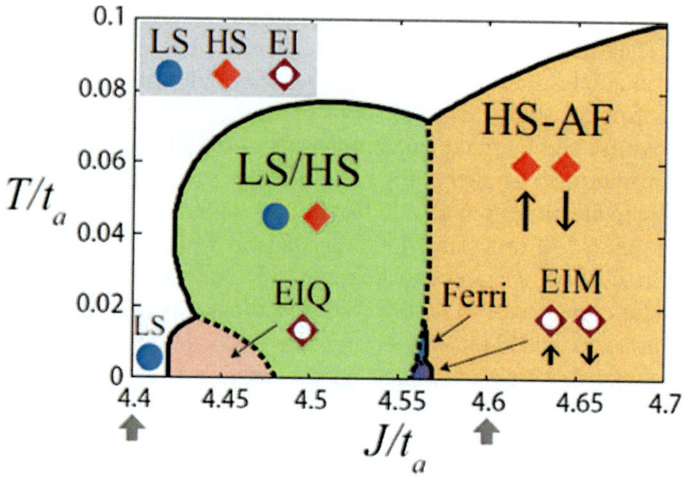

Fig. 3.8 Finite-temperature phase diagram calculated by the effective Hamiltonian. Solid and broken lines represent the second- and first-order phase transitions, respectively. Circles, rhombuses, and their combined symbols represent the LS, HS, and EI states, respectively, and arrows represent schematic spin directions. The ratio of the hopping integrals is chosen to be $t_b/t_a = -0.1$. Data were taken from Ref. [55]

ing temperature, both the EI phases disappear and the LS/HS phase is stabilized in a wide region of J. This is owing to the spin entropy at the HS sites in the LS/HS phase. In order to analyze the magnetic field effect in this system, the Zeeman term is added to the effective Hamiltonian in (3.19), defined as

$$\mathcal{H}_{zeeman} = -H \sum_i S_i^z, \tag{3.23}$$

where the magnetic field H is parallel to the z axis, and the coupling with the orbital angular momentum is neglected. The finite-temperature phase diagram under the magnetic field is shown in Fig. 3.7a. The magnetic field is applied to the LS phase at $J/t_a = 4.4$ as indicated by a bold arrow in Fig. 3.8. In low temperatures, the LS phase is changed into the EIQ phase by applying the magnetic field. The results imply the magnetic field induces the EI phase. With increasing H, sequential phase transitions occur as LS→EIQ→LS/HS→EIQ→HS.

Present results provide a possible interpretation for the experimental observation. The magnetic field $H/t_a = 0.025$ is estimated to be about 100T, and $|t_b/t_a| \sim 0.2$ for the perovskite cobalt oxides. The phase diagrams shown in Figs. 3.7a reproduce qualitatively the experimental phase diagram. There are some qualitative similarities between the theoretical $H - T$ phase diagram and the experimental one shown in Fig. 3.7b. Therefore, we propose a possible interpretation that the B1 and B2 phases are the LS/HS ordered phase and the EIQ phase, respectively, although further investigation and comparisons are necessary to identify the experimentally observed new phases.

3.6 Multi-orbital Hubbard Model

Until now, the studies of the spin-state transition and the EI phase based on the two-orbital Hubbard model and its effective model are introduced. These are the minimal models to investigate the electronic state in the correlated electron system with SSD. On the other hand, in the real system of the transition-metal oxides, such as $LaCoO_3$, a Co^{3+} shows the five $3d$ orbitals which have characteristic anisotropies of the electronic wave function, as well as the hopping integral between the NN sites. In this section, we introduce the theoretical analyses of the Hubbard model where the five $3d$ orbitals are fully taken into account [57–59].

The five-orbital Hubbard model is defined in the same ways in the two-orbital Hubbard model introduced in (3.1) with Eqs. (3.2) and (3.3). In the electron annihilation operators $c_{i\eta\sigma}$, the index η takes the five $3d$ orbitals, and the transfer integrals t_η are determined by the Slater-Koster parameters. The electron number per site is fixed to be 6, corresponding the Co^{3+} ion. The Hartree–Fock approximation was applied to the interaction terms in (3.3), and the numerical calculations were performed in mainly the two-dimensional square lattice, and in the three-dimensional cubic lattice [57].

The calculated phase diagram is presented in the plane of the energy splitting between the e_g and t_{2g} orbitals, Δ, and the Hund coupling J in Fig. 3.9. The results in the cubic lattice are semi-quantitatively the same as this result and are consistent with the results obtained in the two-orbital Hubbard model. In the region of large Δ, the LS phase with the electron configuration $(t_{2g})^6(e_g)^0$ is realized, on the other

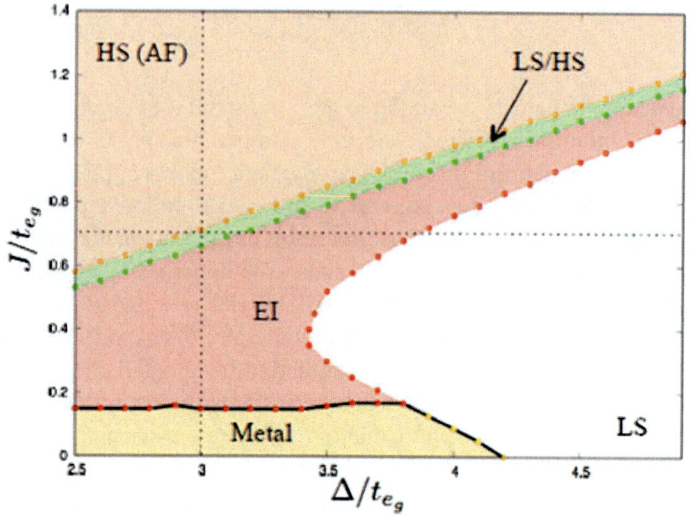

Fig. 3.9 Phase diagram at zero temperature obtained by the five-orbital Hubbard model in a square lattice. Thick lines and thin lines represent the phase boundaries, where the phase transitions are of the first and second orders, respectively. Parameter values are chosen to be $U = 7.5J, U' = U - 2J$, $I = J$ and $t_{t_{2g}} = 0.3t_{e_g}$. Data were taken from Ref. [57]

hand, in the large J region, the HS phase with $(t_{2g})^4(e_g)^2$ appears. In between the LS and HS phases, the LS/HS ordered phase and the EIQ phase appear. The EIM phase seen in Fig. 3.3 is not stabilized in this calculation, which might be attributed to the Hartree–Fock approximation. In addition, the metallic phase is realized in the small Δ and J region.

Details of the order parameters and several physical quantities are plotted as a function of J at $\Delta/t_a = 3$ are shown in Fig. 3.10. The calculated results in the LS and HS phases are expected from their electron configurations. In the HS phase, the d_{xy} is doubly occupied and the other two t_{2g} orbitals are singly occupied. This is due to the two-dimensional square lattice, where the d_{xy} orbital has a larger bandwidth than the other two t_{2g} orbital bands. When we focus on the EI phase, the only component of $\tilde{\tau}^x_{bc} \equiv \sum_\sigma (-1)^\sigma \langle c^\dagger_{x^2-y^2,\sigma} c_{xy,\sigma} + h.c.\rangle$ is finite. We confirm that in the calculation in the three-dimensional square lattice, this component of EI order parameter and other equivalent ones, where the Cartesian coordinates are cyclic permutations, are realized.

In the cubic point group O_h, the two irreducible representations for the EI order are possible, i.e., T_{1g} and T_{2g}. The stabilized EI orders, where the pairs are generated from the electrons in the $d_{x^2-y^2}$ orbital and the holes in the d_{xy} orbital and other equivalent pairs, are classified into T_{1g}. On the other hand, the EI orders, where the pairs are generated from the electrons in the $d_{3z^2-r^2}$ orbital and the holes in the d_{xy} orbital and other equivalent pairs, are classified into T_{2g}. The stability of the EI phase with the T_{1g} irreducible representation is understood from the band structure as follows. In Fig. 3.11, the bare d-orbital bands are shown. The non-interacting bands

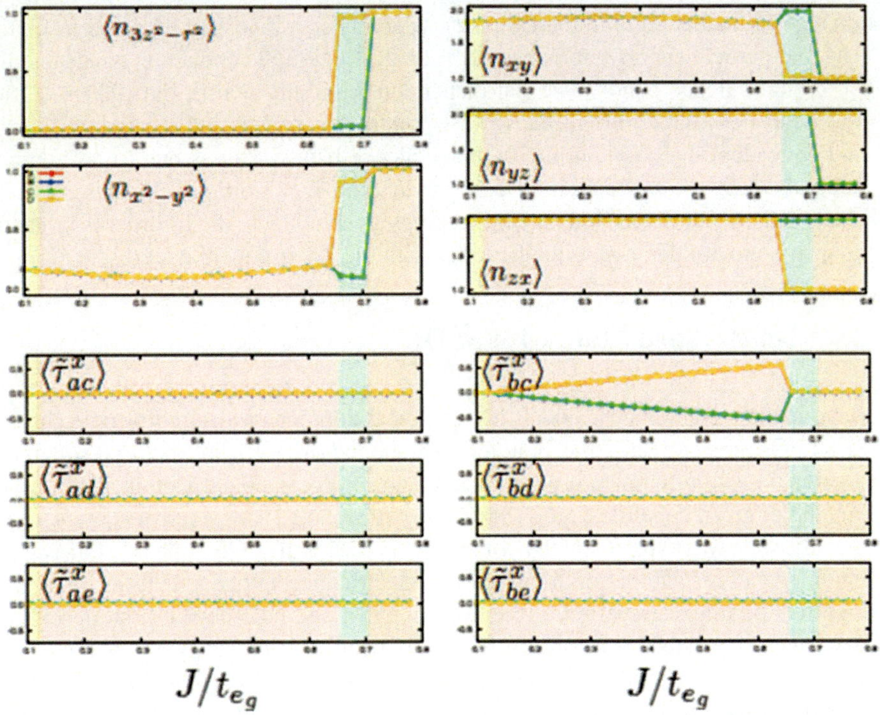

Fig. 3.10 Several physical quantities were obtained by the five-orbital Hubbard model in the square lattice. Averaged electron numbers in orbital $\langle n_\gamma\rangle$, and the EI order parameters $\tilde{\tau}^x_{\alpha\beta}$ where α and β take (a, b, c, d, e) correspond to the orbitals $(d_{3z^2-r^2}, d_{x^2-y^2}, d_{xy}, d_{yz}, d_{zx})$, respectively. Parameter values are chosen to be $U = 7.5J$, $U' = U - 2J$, $I = J$ and $t_{t_{2g}} = 0.3t_{e_g}$. Data were taken from [57]

Fig. 3.11 Non-interacting *d* orbital bands. Data were taken from Ref. [57]

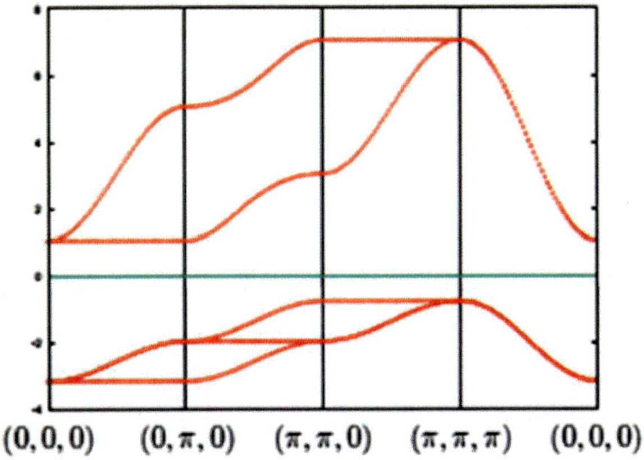

are approximately valid in the LS phase which is classified into the band insulator. The one-particle energy gap opens between the e_g-orbital conduction bands and the t_{2g}-orbital valence bands. The flat-dispersion parts are seen in the (000) – (100) and equivalent directions in the lowest conduction bands, and the (111) – (110) and equivalent directions in the highest valence bands. This is due to no hopping integrals along the z direction for both the d_{xy} and $d_{x^2-y^2}$ orbitals. Because of these characteristic flat-dispersion parts, the energy gain owing to the EI order is larger, in comparison with that due to the EI order with the T_{2g} irreducible representation.

3.7 Hole-Doped State with SSDF

Beyond the stoichiometric cobaltites, chemical dopings of electron or hole carriers in $R\mathrm{CoO}_3$ have attracted much attention since the discovery of the high superconducting transition-temperature cuprates which emerge in doped Mott insulators. The high hole-doped cobaltites, $\mathrm{La}_{1-x}\mathrm{Sr}_x\mathrm{CoO}_3$ with $x > 0.3$–0.4, a ferromagnetic (FM) metallic state was experimentally confirmed. In the lightly hole-doped region, several inhomogeneous features in the magnetic, electric, and lattice structures have been reported. Spatial segregation of the hole-rich FM and hole-poor insulating regions have been suggested by neutron diffraction, electron microscopy, NMR, and other studies [60–68]. The observed giant magneto-resistance is expected to result from electronic and magnetic inhomogeneity [64]. Electronic phase separation has been proposed in several correlated electron systems [69–75].

In this section, we introduce the theoretical analyses of the hole-doped state in the two-orbital Hubbard model and the possibility of the electronic phase separation [50]. The variational Monte-Carlo method is known as one of the powerful methods to analyze the electronic structure in the ground state [76]. The variational wave function is given as the product $\Psi = G|\Phi\rangle$ where G is termed the correlation factor and $|\Phi\rangle$ is the one-body wave function. Two kinds of wave function are adopted in $|\Phi\rangle$. (1) The Slater determinant given as

$$|\Phi_N\rangle = \prod_{\mathbf{k}}{}' \prod_{\gamma\sigma} c^\dagger_{\mathbf{k}\gamma\sigma} |0\rangle , \tag{3.24}$$

where $c_{\mathbf{k}\gamma\sigma}$ is the Fourier transform of $c_{i\gamma\sigma}$, and $\prod'_{\mathbf{k}}$ is a product of $\mathbf{k}$ up to the Fermi momentum. 2) The wave function for the AFM-HS state given as

$$|\Phi_{AF}\rangle = \prod_{\mathbf{k}}{}'' \prod_{\gamma\sigma} d^\dagger_{\mathbf{k}\gamma\sigma} d^\dagger_{\mathbf{k}+\mathbf{K}\gamma\sigma} |0\rangle , \tag{3.25}$$

where $d^\dagger_{\mathbf{k}\gamma\sigma} = \alpha^{(+)}_{\mathbf{k}\gamma} c^\dagger_{\mathbf{k}\gamma\sigma} + \varsigma_\sigma \alpha^{(-)}_{\mathbf{k}\gamma} c^\dagger_{\mathbf{k}+\mathbf{K}\gamma\sigma}$, and $d^\dagger_{\mathbf{k}+\mathbf{K}\gamma\sigma} = -\varsigma_\sigma \alpha^{(-)}_{\mathbf{k}\gamma} c^\dagger_{\mathbf{k}\gamma\sigma} + \alpha^{(+)}_{\mathbf{k}\gamma} c^\dagger_{\mathbf{k}+\mathbf{K}\gamma\sigma}$ where $\mathbf{K}$ is the AFM momentum, and $\prod''_{\mathbf{k}}$ represents a product of $\mathbf{k}$ up to

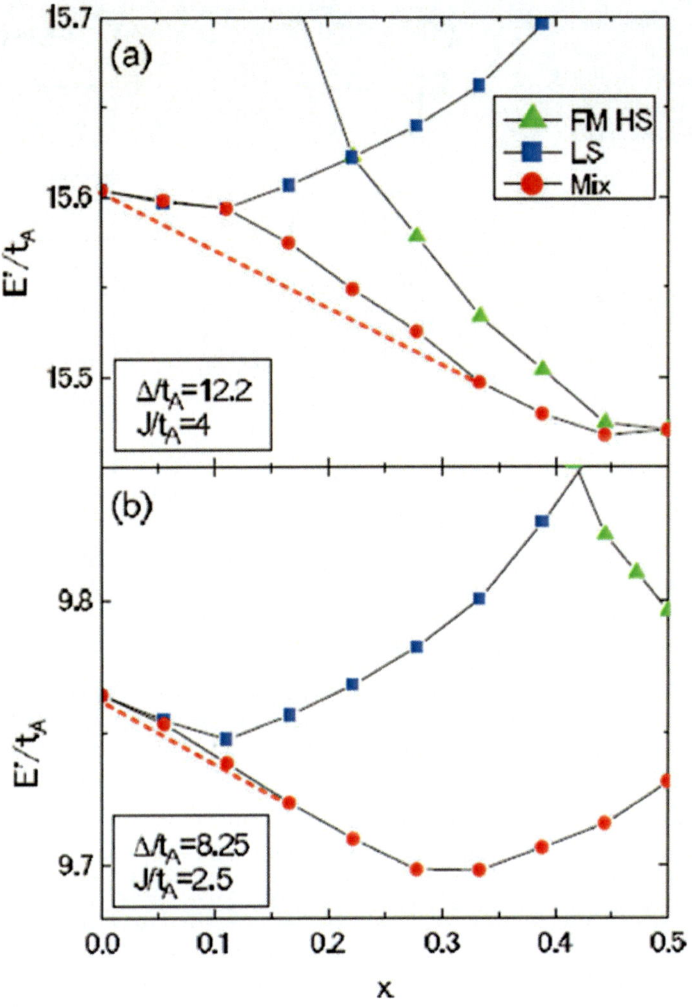

Fig. 3.12 Hole concentration dependences of the energy expectations for several states calculated in the two-orbital Hubbard model at $(\Delta/t_a, J/t_a = 12.2, 4)$ in (**a**), and at $(\Delta/t_a, J/t_a) = (8.25, 2.5)$ in (**b**). The broken lines are given by the Maxwell'fs construction. Data were taken from Ref. [50]

the Fermi momentum in the AFM Brillouin zone. We introduce $\varsigma_{\uparrow(\downarrow)} = 1(-1)$ and the coefficients $\alpha_{\mathbf{k}\gamma}^{(\pm)}$ which are determined from the band dispersion and the variational parameters. As for the Gutzwiller-type correlation factor, it is introduced as $\Pi_{il}(1 - \xi_l \mathscr{P}_{il})$ where l indicates the local electron configurations, $\mathscr{P}_{il}$ is the projection operator at site i for the configuration l, and ξ_l are the variational parameters.

The calculated energies of the several states are shown in Fig. 3.12 as functions of the hole concentration x. Holes are introduced into the LS phase near the phase boundary. The following four states are stabilized: (i) the LS-metallic state where the Fermi surface is located at top of the b (valence) band. (ii) the FM HS-metallic state

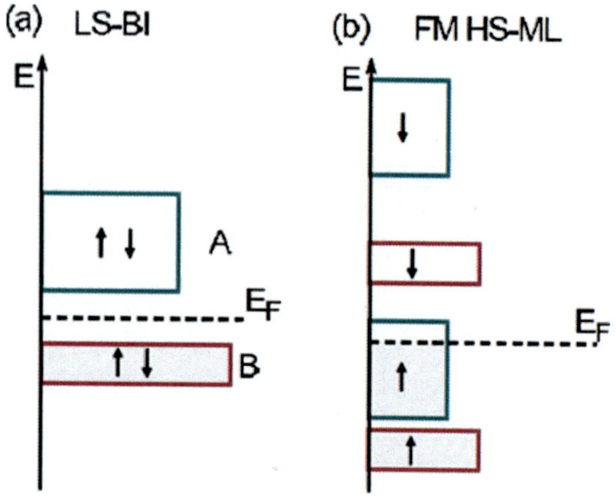

Fig. 3.13 Schematic density of states in the LS state at $x = 0$ and that in the HS-ML state in a high hole-doped region. Data were taken from Ref. [50]

where the Fermi surface is in the a band. (iii) the AFM HS-metallic state where the Fermi surface exists in the a band, and (iv) the mixed state where three kinds of the configurations, $|a^2\rangle$, $|a^1b^1\rangle$ with $S = 1$, and $|b^1\rangle$, are distributed spatially. As shown in Fig. 3.12a, the LS state is destabilized with increasing x. The FM HS-metallic state is realized in a highly doped region. In between the two regions, the mixed state is the lowest energy state. The mixed state is connected to the LS and HS states in the low and high x regions, respectively. The E versus x curve in the mixed state is convex in the region of $0 < x < 0.33$ in Fig. 3.12a and $0 < x < 0.15$ for Fig. 3.12b. By following Maxwell's construction, the phase separation of the LS insulating state and the FM HS dominant mixed states is more stabilized than the homogeneous phase.

Schematic density of states in the LS state at $x = 0$ and the FM HS state in a high hole-doped region are presented in Fig. 3.13. In the LS state at $x = 0$, the Fermi level is located inside the bandgap. The bandwidth of the a band is larger than that of b. On the other hand, in the FM HS-metallic state, the Fermi level is located in the a band. Because of the large bandwidth of the a band, there is a large kinetic energy gain in comparison with the doped LS state where the Fermi level is located in the b band in the rigid-band scheme. The present phase separation phenomena are attributed to this bandwidth difference as follows. In the rigid-band sense, by doping of holes in the LS state, the Fermi level falls into the top of the b band from the middle of the gap as shown in Fig. 3.13a. If we suppose that this state is realized in a low x region and is transferred into the FM HS-metallic state shown in Fig. 3.13b with increasing x, the Fermi level is increased with increasing x. This implies the negative charge compressibility $\kappa = -(\partial \mu / \partial x) < 0$ with the chemical potential.

3.8 Photinduced Spin State in Correlated Systems

The spin-state transition or cross over have been also studied by the optical irradiation, and related photoinduced ultrafast measurements. Some examples have been seen in the insulating organometallic complexes, such as the Prussian blue analog, where magnitudes of the localized spins in Co or Fe ions are switched by photon irradiation [32–39]. Another type of photoinduced spin-state transition is suggested in itinerant correlated electron systems, e.g., the cobalt oxides with a perovskite structure [40–43]. The ultrafast optical measurements in the low-temperature LS insulators show that transient metallic spectra appear within the time scale of 100 femtoseconds, and are distinct from the spectra in the high-temperature phase. In this section, we introduce the theoretical studies of the photoinduced spin-state change and their implication to the experiments [44, 45].

One of the theoretical methods to study the photoinduced state is to construct the effective Hamiltonian where the number of the electron-hole pairs are fixed to be finite value. This is derived from the two-orbital Hubbard model by the perturbations way where the hopping term in Eq. (3.2) is treated as the perturbational Hamiltonian. This is the same way to derive the effective Hamiltonian introduced in Eq. (3.19). In addition to the LS and HS states, the states where the numbers of electrons in a site are one or three [see Fig. 3.14a, b)] are introduced. These local states are termed the hole state and the electron state, respectively. The wave functions are given as $|\psi_{h\sigma}\rangle = c^\dagger_{b\sigma}|0\rangle$, and $|\psi_{e\sigma}\rangle = c^\dagger_{a\sigma}c^\dagger_{b\uparrow}c^\dagger_{b\downarrow}|0\rangle$, respectively. The number of both the electron state and the hole state is one in a N-site cluster. The effective Hamiltonian is classified by the electronic states in the NN sites as

$$\mathscr{H}_1 = \widetilde{\mathscr{H}_0} + \mathscr{H}_{eh} + \mathscr{H}_e + \mathscr{H}_h. \tag{3.26}$$

The first term $\widetilde{\mathscr{H}_0}$ corresponds to $\mathscr{H}_0$ in Eq. (3.19), where neither the electron and hole states are involved in the interactions. The second term describes the interactions between the electron state and the hole state. The third and fourth terms describe the interactions between the electron state and LS or HS, and the interactions between the hole state and LS or HS, respectively. Details of the explicit forms for the Hamiltonian are not shown here and are shown in Ref. [45].

The phase diagram in the photoexcited state calculated in the effective Hamiltonian is shown in Fig. 3.15, together with that in the ground state [45]. The phase

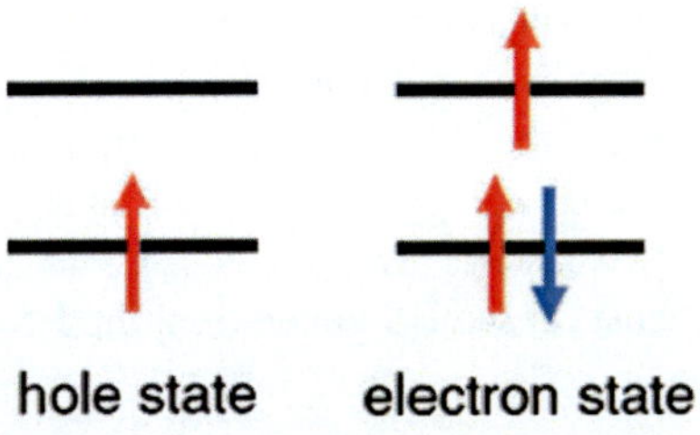

Fig. 3.14 Schematic electron configurations of the hole state and the electron state

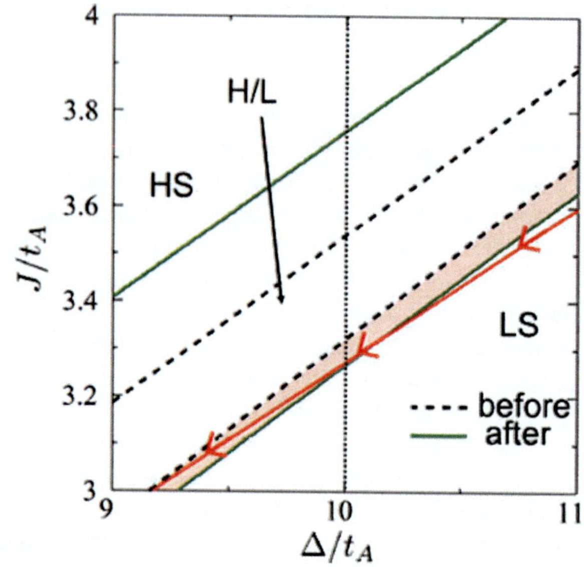

Fig. 3.15 Phase diagram in the $\Delta - J$ plane calculated in the effective Hamiltonian. The broken and bold lines represent the phase boundaries in the ground state and in the photoinduced state, respectively. The parameters are chosen to be $U = 4J$, $U' = 2J$, and $t_b = 0.05t_a$. Data were taken from Ref. [45]

boundary between the LS and LS-HS mixed phases shifts to a region of the LS phase. There is a parameter region where the LS phase in the ground state is changed into the LS-HS mixed phase in the photoexcited state. That is, the photoirradiation induces the HS state in the LS phase at the vicinity of the phase boundary. The electronic structure in the photoinduced HS state was examined. The electronic-state distribution functions are defined by $g^M(n) = z^{-1} \sum_{j \in nNN} \sum_i \langle P^M_{i+j} P^h_i \rangle$ where $\sum_{j \in nNN}$ implies a summation of j connecting the nth NN sites of i, and z is the number of the nth NN sites. The operator P_i^M $(M = L, H, e)$ is the projection operator for the local electronic structure. This function describes the distribution of the local electronic states at the nth NN sites from the photoinduced hole state. The calculated results of the distribution functions in a two-dimensional cluster are shown in Fig. 3.16b. It is shown that $g^H(n)$ is nearly 0.25 at $n = 1$ and zero at $n > 2$. This implies a local bound state between the HS state and a photodoped hole state. The results in the one-dimensional system are shown in the inset of Fig. 3.16b. The HS distribution is located at the NN sites of the hole state.

The mechanism of the FM HS-hole bound state is the following. Let us consider the case of $I = 0$ for simplicity. In the ground state, the energy difference per site between the LS and HS states is $E_{HS} - E_{LS} = \Delta - J$ in the local limit. When the HS-hole bound state is not generated, the kinetic energy of the hole state is $-zt_b$, and is the exchange constant between the hole state and the LS state t_b. On the other hand, when the HS-hole bound state is generated, the energy gain is given by the bonding orbital energy in the bound state as $-t_a$. Then, the energy difference between the two cases is $\Delta E_{BS} = (-t_a) - (-zt_b)$. When this energy gain overcomes the energy cost for the HS generation, the HS-hole bound state is realized.

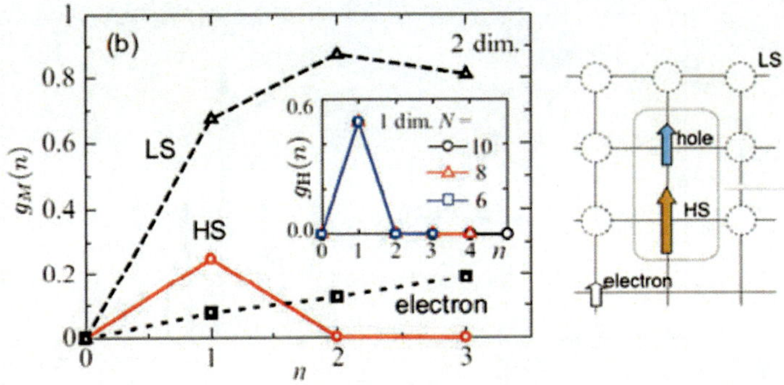

Fig. 3.16 Distribution function of the electronic states as a function of the distance from the hole state. The bold, broken, and dotted lines represent the distribution functions for the HS, LS, and electron states, respectively. The inset shows distribution functions for the HS state in one-dimensional clusters. The parameters are chosen to be $J = 3.3t_a$, $U = 4J$, $U' = 2J$, $\Delta = 10t_a$, and $t_b = 0.05t_a$. Right shows a schematic picture for the hole-HS bound state. Data were taken from Ref. [45]

Another method to analyze the photoexcited electronic state is to calculate the real-time evolution of the electronic structure after photoirradiation. Here we introduce the results by using the time-dependent Hartree–Fock approximation applied to the two-orbital Hubbard model [45]. The photoirradiation is simulated by excitations of electrons from the highest-occupied levels to the lowest-unoccupied ones at time $\tau = 0$. The time evolution of the wave function is calculated in the time-dependent Shrödinger equation for the ν-th level, $|\phi_\nu(\tau)\rangle$ given as [77, 78],

$$|\phi_\nu(\tau)\rangle = P \exp\left[-i \int_0^\tau d\tau' \mathscr{H}_{\rm MF}(\tau')\right] |\phi_\nu(0)\rangle, \tag{3.27}$$

where $\mathscr{H}_{\rm MF}(\tau)$ is the time-dependent mean-field Hamiltonian, and P is the time-ordering operator. The wave function at time $\tau + d\tau$, with a short time increment $d\tau$, is calculated from the wave function at time τ by expanding the exponential factor as

$$|\phi_\nu(\tau + d\tau)\rangle = \sum_\mu \langle\varphi_\mu(\tau)|\phi_\nu(\tau)\rangle e^{-i\varepsilon_\mu(\tau)d\tau} |\varphi_\mu(\tau)\rangle, \tag{3.28}$$

where $|\varphi_\mu(\tau)\rangle$ is the eigen state of $\mathscr{H}_{\rm MF}(\tau)$ with the eigen energy $\varepsilon_\mu(\tau)$.

The numerical results of the time-dependent density of states are shown in Fig. 3.17 [45]. At $\tau = 10t_a^{-1}$, an energy gap exists between the narrow b band and the wide a band. Tiny weights of the hole and electron parts of the density of states are observed at the top of the b band and the bottom of the a band, respectively. After photoirradiation, the top of the valence band and the bottom of the conduction band start to separate from the main bands. Finally, the original gap is almost filled by

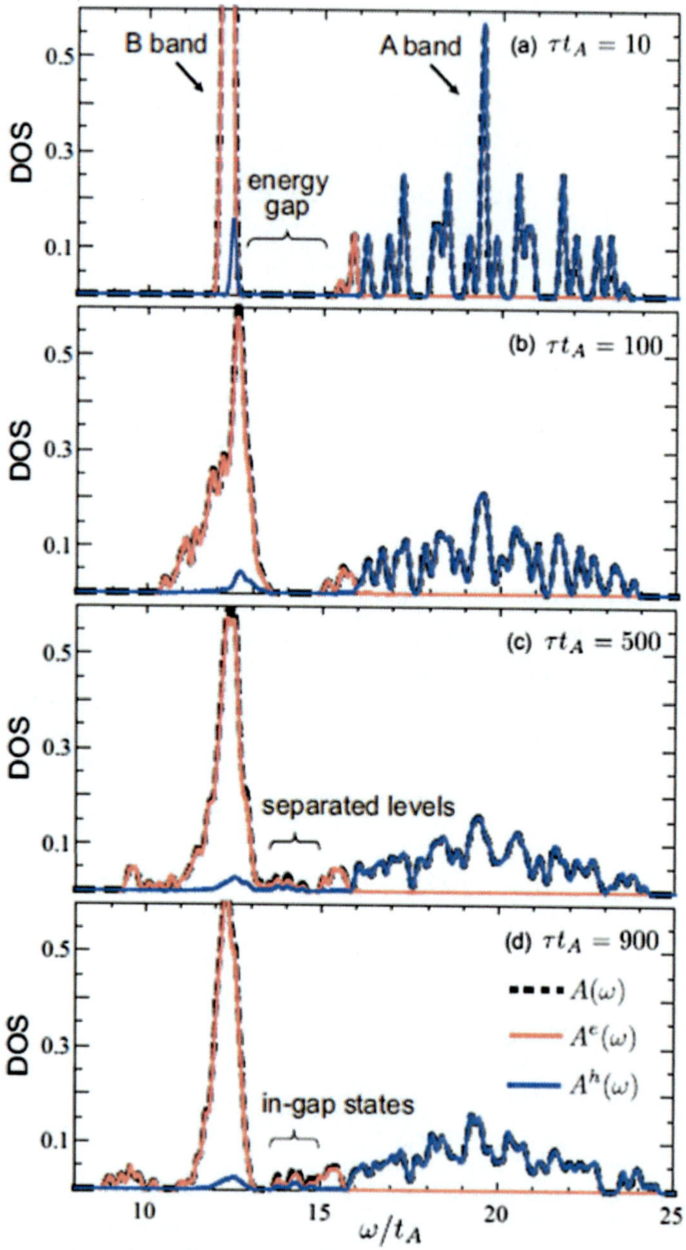

Fig. 3.17 Time dependences of the density of states. The electron part and the hole part are represented by pink solid and blue solid lines, respectively. The parameters are chosen to be $U = 4J$, $U' = 2J$, $J = 3.1t_a$, $\Delta = 10t_a$, $t_b = 0.05t_a$. Data were taken from Ref. [45]

in-gap states originating from the localization of the electron and hole states. These data are consistent with the localization of the electron and hole states mentioned above.

3.9 Summary and Perspective

In this article, we review the theoretical studies on the correlated electron systems with the SSDF, in particular, the cobaltates with the perovskite structure. The competition between the crystalline field splitting and the Hund coupling controls the stability of the LS state with $S = 0$ and the HS state with $S = 2$. Under the competition, the IS state with $S = 1$ has been proposed as a possible hidden spin-state long time ago. We introduce the following three electronic states as other plausible exotic hidden states:

(i) Excitonic insulating state The quantum mechanical mixing of the HS and LS states can reduce the electronic energy due to the quantum hybridization. This is a Z_2 symmetry broken state associated with lowering a local point-group symmetry. The two kinds of the EI phase are possibly stabilized: the EIQ and EIM phase in which the weights of the LS and HS are dominant, respectively. The longitudinal spin excitation attributed to the Higgs excitation mode of the EI order parameter can be utilized to identify the EI phase. The recent discovered novel phase induced by the high magnetic field has a possibility of the field-induced EI phase, where the HS component in the wave function contributes to the Zeeman energy gain. The characteristic dispersion relations in the 3*d* bands provide large energy gain by the EI order with the T_{1g} irreducible representation.

(ii) Electronic phase separation A phase separated state between the LS band insulating state and the FM metallic state, i.e., a doped HS Mott insulating state, has a chance to be realized by hole-carrier doping. This is due to the difference between the bandwidths in the two phases; the larger (smaller) in the FM metallic state (the LS band insulating state). In actual systems, the nano-scale phase inhomogeneity may occur due to the Coulomb interaction between the hole carriers in the hole-rich FM metallic state. This theoretical prediction is consistent with the several experimental results suggesting the inhomogeneity in electronic, magnetic, and structural states, and explains the giant megnetoresistance observed in lightly doped cobaltates.

(iii) Photoinduced HS-hole bound state A bound state between the HS state and the photoinjected hole state is generated by photoirradiation in the LS band insulating state. This state originates from the kinetic energy gain of the photoinjected hole which can be mobile in the case that the LS state is changed into the HS state. This is identified as a photoinduced HS polaron state associated with the spin crossover and can explain the systematic measurements of the photoinduced ingap state obtained by the optical pump-probe experiments in the layered cobaltates.

In order to confirm the hidden exotic electronic states proposed by the theoretical calculations, further intensive and extensive investigations are required. One essential subject which was not touched in this article is the lattice effect. Since a Co ion with the HS state has a larger ionic radius than that with the LS state, the volume expansion of the O_6 octahedral and the cooperative elastic distortion needs to be taken into account to discuss the spin-state transition/cross over. There are the orbital degrees of freedom in the HS and LS state which induces the Jahn-Teller type lattice distortions. Not only the electron correlation and itinerant effect, but also the lattice distortion and the electron-lattice interaction effects should be taken into account in the electronic model and analyses. This might be essential to investigate the photoinduced transient state where the different energy scales of the electron and lattice systems provide the different time scales. There is a chance to identify the dominant interactions between the local SSDFs by using the ultrafast optical measurements.

Acknowledgements The authors would like to thank J. Nasu, M. Naka, T. Watanabe, J. Ohara, Y. Kanamori, R. Suzuki, T. Tatsuno, E. Mizoguchi, A. Ono, Y. Masaki, S. Koshihara, S. Iwai, H. Okamoto, and Y. Okimoto for their fruitful discussions. This work was supported by JSPS KAKENHI, Grant Numbers JP17H02916 and JP18H05208. Some of the numerical calculations were performed using the facilities of the Supercomputer Center, the Institute for Solid State Physics, The University of Tokyo.

References

1. S. Maekawa, T. Tohyama, S. Barnes, S. Ishihara, W. Koshibae, G. Khalliulin, *Physics of Transition Metal Oxides* (Springer Verlag, Berlin, 2004). and references therein
2. M. Imada, A. Fujimori, Y. Tokura, Rev. Mod. Phys. **70**, 1039 (1998)
3. H. Gretarsson, S. R. Saha, T. Drye, J. Paglione, J. Kim, D. Casa, T. Gog, W. Wu, S. R. Julian, and Y.-J. Kim, Phys. Rev. Lett.**110**0470032013
4. G. J. Halder, C. J. Kepert, B. Moubaraki, K. S. Murray, and J. D. Cashion, Science **298**17622002
5. J.-F. Lin, V. V Struzhkin, S. D. Jacobsen, M. Y. Hu, P. Chow, J. Kung, H. Liu, H.-K. Mao, and R. J. Hemley, Nature**436**3772005
6. D. Antonangeli, J. Siebert, C. M. Aracne, D. L. Farber, A. Bosak, M. Hoesch, M. Krisch, F. J. Ryerson, G. Fiquet, and J. Badro, Science**331**642011
7. H. Hsu, P. Blaha, M. Cococcioni, and R. M. Wentzcovitch, Phys. Rev. Lett.**106**1185012011
8. K. Asai, P. Gehring, H. Chou, G. Shirane, Phys. Rev. B **40**, 10982 (1989)
9. Y. Tokura, Y. Okimoto, S. Yamaguchi, H. Taniguchi, T. Kimura, H. Takagi, Phys. Rev. B **58**, 1699(R) (1998)
10. R. Mahendiran and A. K. Raychaudhuri, Phys. Rev. B **54**160441996
11. K. Sato, A. Matsuo, K. Kindo, Y. Kobayashi, and K. Asai, J. Phys. Soc. Jpn. **78**0937022009
12. J. Yu, D. Louca, D. Phelan, K. Tomiyasu, K. Horigane, and K. Yamada, Phys. Rev. B **80**0524022009
13. G. Vankó, J.-P. Rueff, A. Mattila, Z. Németh, and A. Shukla, Phys. Rev. B **73**0244242006
14. Y. Tanabe, S. Sugano, J. Phys. Soc. Jpn. **9**, 766 (1954)
15. S. Yamaguchi Y. Okimoto, H. Taniguchi, and Y. Tokura, Phys. Rev. B **53**, R2926 (1996)
16. M.A. Korotin, S.Y. Ezhov, I.V. Solovyev, V.I. Anisimov, D.I. Khomskii, G.A. Sawatzky, Phys. Rev. B **54**, 5309 (1996)
17. M.W. Haverkort, Z. Hu, J.C. Cezar, T. Burnus, H. Hartmann, M. Reuther, C. Zobel, T. Lorenz, A. Tanaka, N.B. Brookes, H.H. Hsieh, H.-J. Lin, C.T. Chen, L.H. Tjeng, Phys. Rev. Lett. **97**, 176405 (2006)

18. S. Noguchi, S. Kawamata, K. Okuda, H. Nojiri, M. Motokawa, Phys. Rev. B **66**, 094404 (2002)
19. Y. Kobayashi, T.S. Naing, M. Suzuki, M. Akimitsu, K. Asai, K. Yamada, J. Akimitsu, P. Manuel, J.M. Tranquada, G. Shirane, Phys. Rev. B **72**, 174405 (2005)
20. D.I. Khomskii, U. Löw, Phys. Rev. B **69**, 184401 (2004)
21. J. Fujioka, Y. Yamasaki, A. Doi, H. Nakao, R. Kumai, Y. Murakami, M. Nakamura, M. Kawasaki, T. Arima, and Y. Tokura, Phys. Rev. B **92**1951152015
22. N. F. Mott, Philos. Mag.**6**2871961
23. D. Jérome, T. M. Rice, and W. Kohn, Phys. Rev.**158**4621967
24. B. I. Halperin and T. M. Rice, Rev. Mod. Phys. **40**7551968
25. H. Fukuyama and T. Nagai, J. Phys. Soc. Jpn. **31**8121971
26. Y. Kuramoto and M. Morimoto, J. Phys. Soc. Jpn. **44**17591978
27. Y. Wakisaka, T. Sudayama, K. Takubo, T. Mizokawa, M. Arita, H. Namatame, M. Taniguchi, N. Katayama, M. Nohara, H. Takagi, Phys. Rev. Lett. **103**, 026402 (2009)
28. T. Kaneko, T. Toriyama, T. Konishi, Y. Ohta, Phys. Rev. B **87**, 035121 (2013)
29. J. Kuněs and P. Augustinský, Phys. Rev. B **89**1151342014
30. J. Kuněs and P. Augustinský, Phys. Rev. B **90**2351122014
31. J. Nasu, T. Watanabe, M. Naka, and S. Ishihara, Phys. Rev. B **93**2051362016
32. O. Sato, T. Iyoda, A. Fukushima, K. Hashimoto, Science **272**, 704 (1996)
33. A. Bleuzen, C. Lomenech, V. Escax, F. Villain, F. Varret, C. C. dit Moulin, and M. Verdaguer, J. Am. Chem. Soc. **122** 6648 (2000)
34. V. Escax, A. Bleuzen, C. C. dit Moulin, F. Villain, A. Goujon, F. Varret, and M. Verdaguer, J. Am. Chem. Soc. **123** 12536 (2001)
35. O. Sato, J. Photoch. Photobio. C: Photochem. Rev. **5**, 203 (2004)
36. N. Willenbacher, H. Spiering, J. Phys. C **21**, 1423 (1988)
37. A.L. Tchougréeff, M.B. Darkhovskii, Int. J. Quantum Chem. **57**, 903 (1996)
38. M. Nishino, K. Boukheddaden, Y. Konishi, S. Miyashita, Phys. Rev. Lett. **98**, 247203 (2007)
39. S. Miyashita, P.A. Rikvold, T. Mori, Y. Konishi, M. Nishino, H. Tokoro, Phys. Rev. B **80**, 064414 (2009)
40. Y. Okimoto, X. Peng, M. Tamura, T. Morita, K. Onda, T. Ishikawa, S. Koshihara, N. Todoroki, T. Kyomen, M. Itoh, Phys. Rev. Lett. **103**, 027402 (2009)
41. Y. Okimoto, T. Miyata, M.S. Endo, M. Kurashima, K. Onda, T. Ishikawa, S. Koshihara, M. Lorenc, E. Collet, H. Cailleau, T. Arima, Phys. Rev. B **84**, 121102(R) (2011)
42. Y. Okimoto, T. Egawa, R. Fukaya, Y. Matsubara, Y. Yamada, N. Yamaya, T. Ishikawa, K. Onda, S. Koshihara, H. Taniguchi, M. Itoh, A. Isayama, and T. Sasagawa, J. Phys. Soc. Jpn. **82**0747212013
43. T. Ishikawa, K. Yamada, H. Itoh, S. Iwai, T. Arima, S. Yamada, and T. Sasaki, EPJ Web Conf. **41**030132013
44. Y. Kanamori, H. Matsueda, and S. Ishihara, Phys. Rev. Lett.**107**1674032011
45. Y. Kanamori, J. Ohara, and S. Ishihara, Phys. Rev. B **86**0451372012
46. P. Werner, A.J. Millis, Phys. Rev. Lett. **99**, 126405 (2007)
47. K. Sano, Y. Ono, J. Phys. Soc. Jpn. **72**, 1847 (2003)
48. K. Kobayashi, H. Yokoyama, Physica C **445–448**, 162 (2006)
49. K. Kubo, Phys. Rev. B **79**, 020407 (2009)
50. R. Suzuki, T. Watanabe, and S. Ishihara, Phys. Rev. B **80**0544102009
51. M. M. Altarawneh, G. W. Chern, N. Harrison, C. D. Batista, A. Uchida, M. Jaime, D. G. Rickel, S. A. Crooker, C. H. Mielke, J. B. Betts, J. F. Mitchell, and M. J. R. Hoch, Phys. Rev. Lett.**109**0372012012
52. V. V. Platonov, Y. B. Kudasov, M. P. Monakhov, and O. M. Tatsenko, Phys. Solid State**54**272012
53. M. Rotter, Z. S. Wang, A. T. Boothroyd, D. Prabhakaran, A. Tanaka, and M. Doerr, Sci. Rep. **4**70032014
54. A. Ikeda, T. Nomura, S. Takeyama, Y.H. Matsuda, A. Matsuo, K. Kindo, K. Sato, Phys. Rev. B **93**, 220401(R) (2016)
55. T. Tatsuno, E. Mizoguchi, J. Nasu, M. Naka, S. Ishihara, J. Phys. Soc. Jpn. **85**, 083706 (2016)
56. A. Sotnikov, J. Kuněs, Sci. Rep. **6**, 30510 (2016)

57. T. Tatsuno, Master thesis in Tohoku University 2018
58. J. Fernandez Afonso and J. Kunes, Phys. Rev. B **95** 115131 (2017)
59. T. Yamaguchi, K. Sugimoto, and Y. Ohta J. Phys. Soc. Jpn. **86**, 043701 (2017)
60. M. Itoh, I. Natori, S. Kubota, K. Motoya, J. Phys. Soc. Jpn. **63**, 1486 (1994)
61. R. Caciuffo, D. Rinaldi, G. Barucca, J. Mira J. Rivas, M. A. Senaris-Rodriguez, P. G. Radaelli, D. Fiorani, and J. B. Goodenough, Phys. Rev. B **59**, 1068 (1999)
62. P.L. Kuhns, M.J.R. Hoch, W.G. Moulton, A.P. Reyes, J. Wu, C. Leighton, Phys. Rev. Lett. **91**, 127202 (2003)
63. A. Ghoshray, B. Bandyopadhyay, K. Ghoshray, V. Morchshakov K. Bärner, I. O. Troyanchuk, H. Nakamura, T. Kohara, G. Y. Liu and G. H. Rao, Phys. Rev. B **69**, 064424 (2004)
64. J. Wu, J.W. Lynn, C.J. Glinka, J. Burley, H. Zheng, J.F. Mitchell, C. Leighton, Phys. Rev. Lett. **94**, 037201 (2005)
65. D. Phelan, Despina Louca, S. Rosenkranz, S.-H. Lee, Y. Qiu, P. J. Chupas, R. Osborn, H. Zheng, J. F. Mitchell, J. R. D. Copley, J. L. Sarrao, and Y. Moritomo, Phys. Rev. Lett. **96**, 027201 (2006)
66. D. Phelan, Despina Louca, K. Kamazawa, S.-H. Lee, S. N. Ancona, S. Rosenkranz, Y. Motome, M. F. Hundley, J. F. Mitchell, and Y. Moritomo, Phys. Rev. Lett. **97**, 235501 (2006)
67. C. Frontera, J.L. García-Muñoz, A. Llobet, M.A.G. Aranda, Phys. Rev. B **65**, 180405(R) (2002)
68. S. Tsubouchi, T. Kyomen, M. Itoh, P. Ganguly, M. Oguni, Y. Shimojo, Y. Morii, Y. Ishii, Phys. Rev. B **66**, 052418 (2002)
69. For reviews, see for examples, S. A. Kivelson, I. P. Bindloss, E. Fradkin, V. Oganesyan, J. M. Tranquada, A. Kapitulnik, and C. Howald, Rev. Mod. Phys. **75**, 1201 (2003), and E. Dagotto, *The Physics of Manganites and Related Compounds*, (Springer-Verlag, Berlin 2003)
70. V.J. Emery, S.A. Kivelson, H.Q. Lin, Phys. Rev. Lett. **64**, 475 (1990)
71. E.L. Nagaev, Phys. State. Sol. (b) **186**, 9 (1994)
72. S. Okamoto, S. Ishihara, S. Maekawa, Phys. Rev. B **61**, 451 (2000)
73. K.I. Kugel et al., Phys. Rev. Lett. **95**, 267210 (2005)
74. M. Lugas, L. Spanu, F. Becca, S. Sorella, Phys. Rev. B **74**, 165122 (2006)
75. A.O. Sboychakov, K.I. Kugel, A.L. Rakhmanov, D.I. Khomskii, Phys. Rev. B **80**, 024423 (2009)
76. C.J. Umrigar, K.G. Wilson, J.W. Wilkins, Phys. Rev. Lett **60**, 1719 (1988)
77. A.D. McLachlan, M.A. Ball, Rev. Mod. Phys. **36**, 844 (1964)
78. A. Terai, Y. Ono, Prog. Theor. Phys. Suppl. **113**, 177 (1993)

Chapter 4
Photoexcited State and Ultrafast Dynamics in Spin-Crossover Cobalt Oxides

Yoichi Okimoto, Tadahiko Ishikawa, and Shin-ya Koshihara

Abstract One of the most interesting topics regarding spin-crossover (SC) materials is optical control of the electronic state. In this chapter, recent trials on the photocontrol of some SC cobaltites are reviewed. In contrast to iron SC complexes, the time scale required for photonic change is on the sub-picosecond level for the cobaltites in general, so femtosecond laser spectroscopy is indispensable for revealing photoinduced SC phenomena. On the basis of an ultrafast linear and nonlinear spectroscopic technique using femtosecond laser pulses, we discuss the photoinduced change of the electronic structure as well as real space dynamics for some exotic SC cobaltites, $Pr_{0.5}Ca_{0.5}CoO_3$ and $BiCoO_3$.

Keywords FS laser pulse · Photoinduced phase transition · Time-resolved spectroscopy · Second harmonic generation

4.1 Introduction

Phase transition phenomena in solids are an important research theme for solid-state physics and statistical physics. The phase that is the subject of such research is often a static state. In recent years, however, much attention has been paid to research that investigates the dynamic electronic phase and its dynamics in an excited state inherent to the material with light irradiation. Phase change driven by light is called "photoinduced phase transition" and has been demonstrated for a quarter of a century for many solid materials such as charge transfer complexes, transition-metal oxides, and spin-crossover (SC) Fe complexes [1]. The research field of this photoinduced phase transition phenomenon has recently expanded and been especially stimulated by the development of fs laser light sources over the past decade. The fs laser pulse has a pulse width of approximately 100 fs and a high photon density per pulse, which makes it possible not only to create new excited states concealed in a solid

Y. Okimoto (✉) · T. Ishikawa · S. Koshihara
Department of Chemistry, Tokyo Institute of Technology,
O-okayama, Tokyo 152-8551, Japan
e-mail: okimoto.y.aa@m.titech.ac.jp

Y. Okimoto et al. (eds.), *Spin-Crossover Cobaltite*, Springer Series
in Materials Science 305, https://doi.org/10.1007/978-981-15-7929-5_4

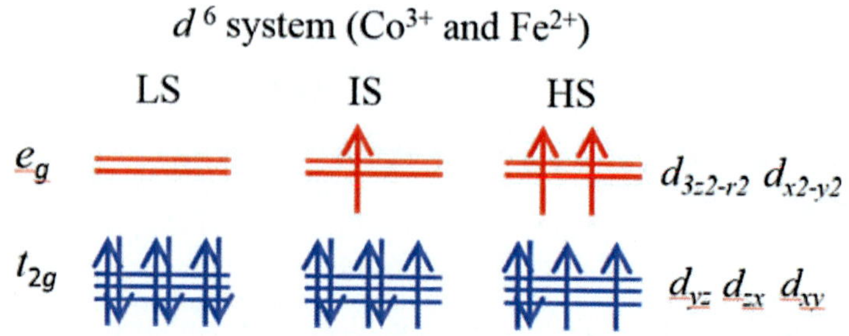

Fig. 4.1 Possible spin configurations of d electrons in Co^{3+}; low spin (LS), intermediate spin (IS), and high spin (HS)

that cannot be detected with lamp light or even an ns laser beam but also to track their time evolution on the time scale of fs. Recently, fs pulses stable from $\approx$ 10 fs to sub-fs can be fabricated, and new techniques that use ultrashort light sources have increasingly been applied in many research fields. Thus, research on photoinduced phase transition is entering a new phase in which materials science and light source work are being developed together.

Among the many examples showing photoinduced transition, there are two important systems; one is strongly correlated electron materials with a perovskite-type structure. In addition to stable and robust chemical properties, the system shows various interesting electric and magnetic phenomena caused by d electrons in transition-metal cations [2]. In some correlated materials, importantly, those electronic properties can also be controlled by laser light irradiation [3, 4], providing a significant stage for photoinduced phase transition research.

The other is comprised of SC complexes containing divalent iron ions (Fe^{2+}), and the illumination of light shows the SC change between a low spin (LS; t_{2g}^6) and high spin [(HS; $e_g^2 t_{2g}^4$)] [5] as shown in Fig. 4.1. Photoinduced SC change, which is often called "light induced excited spin state trapping" (LIESST), can reversibly occur between the HS and LS state depending on the wavelength of excitation light. In addition, the real space dynamics [9] and structural dynamics [10] concerning the LIESST phenomena are also reported.

In this book, we focus on the unique SC behavior in perovskite-type Co oxides [6–8]. As described in detail in other chapters, SC change can also be realized by applying various external stimuli, such as changing the temperature, applying a strong magnetic field, and so on. Some of the results are different from those seen in iron complexes, because the cobalt system has not only the SC aspect but also a strong correlation between spins. In this chapter, we notice some Co perovskite systems exhibiting an SC phenomenon, and we use an fs laser technique to reveal the photoinduced phase change and ultrafast dynamics characteristic to cobalt systems.

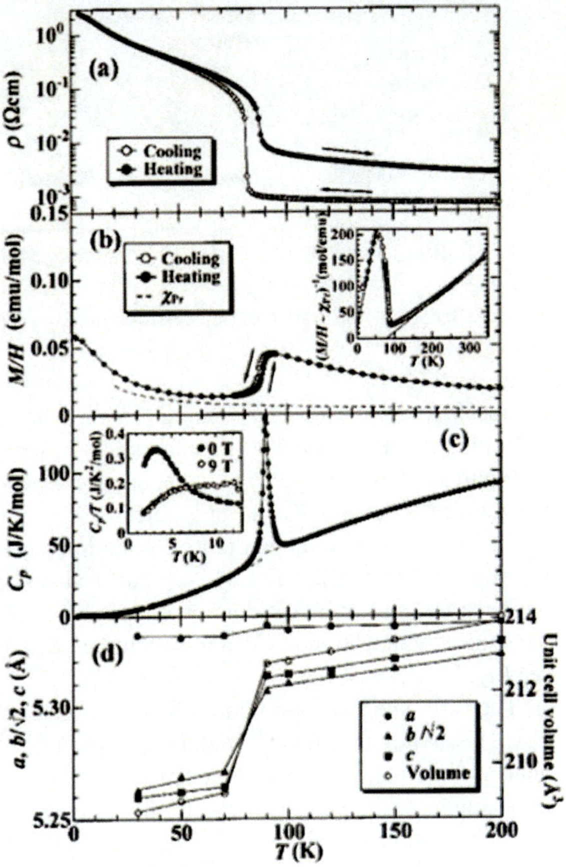

Fig. 4.2 Temperature dependence of resistivity (**a**), magnetization (**b**), heat capacity (**c**), and lattice constants and unit cell volume in $Pr_{0.5}Ca_{0.5}CoO_3$. Reprinted with permission from Ref. [13] ©(2002) by the American Physical Society

4.2 Photoexcited State in $Pr_{0.5}Ca_{0.5}CoO_3$

4.2.1 Spin-Crossover and Insulator-Metal Transition in Cobaltite

The first example we introduce is $Pr_{0.5}Ca_{0.5}CoO_3$. Before discussing the photoexcited state, let us briefly confirm the spin state concerning the SC transition. Figure 4.1 shows possible spin configurations in a d^6 system (e.g., Co^{3+} or Fe^{2+}). When the crystal field is large, the LS state is observed. However, Hund's rule coupling is greater than the crystal field, an intermediate spin (IS; $t_{2g}^5 e_g^1$) or HS state appears. It is said that the IS state is stabilized by the dynamical Jahn-Teller effect as demonstrated in several works in the Co^{3+} system [11, 12]. This is in contrast to other SC-system iron complexes containing Fe^{2+}, for which the IS state has scarcely been reported.

Among many cobalt oxides, $Pr_{0.5}Ca_{0.5}CoO_3$ shows an unconventional insulator-metal transition at around $T_{IM} \approx 89$ K, which was reported by Itoh and his coworkers

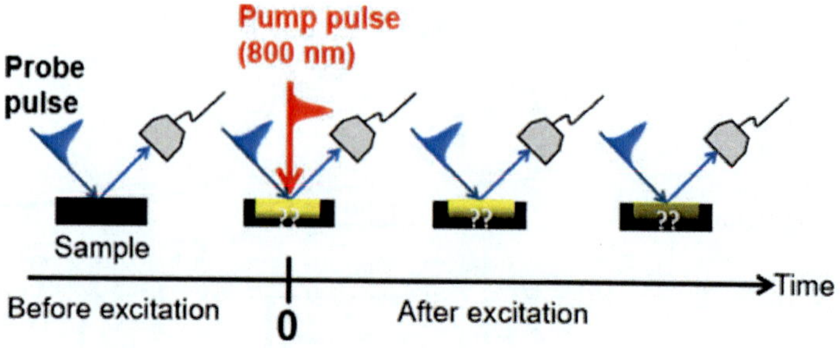

Fig. 4.3 Schematic illustration of time resolved pump-probe reflection spectroscopy

for the first time [13]. The nominal valence of Co is +3.5 and Co^{3+}:Co^{4+}=1:1, indicating that $Pr_{0.5}Ca_{0.5}CoO_3$ contains Co^{3+} ions.

Figures 4.2a and b display the temperature dependence of resistivity (ρ) and magnetization (M) in $Pr_{0.5}Ca_{0.5}CoO_3$ after [13]. With a decreasing temperature, the $\rho - T$ curve shows a sudden jump from a low resistive metallic state to an insulating state at around T_{IM}. The $M - T$ curve also shows a sharp change at the same temperature, indicating a reduction of spin below T_{IM}. These jumps are assigned as the spin state transition in Co^{3+} between the LS state ($S = 0$) to the IS state ($S = 1$, see Fig. 4.1).

The Co^{4+} site is always in the LS state, which causes a Curie-like increase below T_{IM}. This assignment of the spin state is supported by the results of recent specific heat analyses [13, 14], nuclear magnetic resonance [15], and photo-emission spectroscopy [16] studies, although some exotic ideas are proposed for the transition, such as valence instability of Pr^{3+} site [17] or an excitonic gap formation at T_{IM} discussed in [18]. (As for the possible excitonic feature in cobaltites, see also Chap. 3.)

An important aspect of the SC phenomenon in $Pr_{0.5}Ca_{0.5}CoO_3$ is that a higher spin state above T_{IM} is relatively conductive driven by itinerant excited e_g carriers, which is different from the case of iron SC complexes. Thus, $Pr_{0.5}Ca_{0.5}CoO_3$ can be viewed as a strongly correlated SC system.

Hereafter, we show the results of ultrafast spectroscopy for $Pr_{0.5}Ca_{0.5}CoO_3$ done using a femtosecond laser system and demonstrate ultrafast photoinduced phase change from an LS insulating state to an IS metallic state [19].

4.2.2 *Experiment*

For the measurement of the photoexcited state, we prepared polycrystalline samples of $Pr_{0.5}Ca_{0.5}CoO_3$, and details on the fabrication are given in [13, 14].

The relative change of reflectivity ($\Delta R/R$) and its time dependence caused by femtosecond laser pulse irradiation was obtained by using the conventional pump-probe method and a Ti:sapphire regenerative amplified laser (pulse width of $\approx$ 120 fs, repetition rate of 1 kHz, photon energy of $\approx$ 1.6 eV) as a light source. The amplified light was divided into two beams; one was used as pump light for the excitation, and the other as probe light after the frequency was converted from 0.5 to 2.1 eV with an optical parametric amplifier (OPA). All the pump-probe spectroscopy was done at 30 K using a He flow-type cryostat. The concept of the pump-probe reflection spectroscopy we conducted is illustrated in Fig. 4.3, which enables us to see how the electronic state of the material varies after the photoexcitation (i.e., the dynamics of the photoexcited state).

4.2.3 *Photoexcitation and Ultrafast Dynamics of* $Pr_{0.5}Ca_{0.5}CoO_3$

First, let us explain the electronic structure of $Pr_{0.5}Ca_{0.5}CoO_3$. At 290 K, according to the literature [19], optical conductivity spectrum ($\sigma(\omega)$) which corresponds to an absorption spectrum has a large spectral weight in the mid-infrared region, indicating the conductive electronic state. At 30 K, in contrast, the spectral intensity of $\sigma(\omega)$ around the mid-infrared region is remarkably suppressed, reflecting the gap opening with the insulator-metal transition. At 30 K, $\sigma(\omega)$ showed two peaks [19], at $\approx$ 0.9 eV and $\approx$ 1.9 eV. The two peaks can be assigned as the charge-transfer transitions from the O $2p$ band to the empty Co t_{2g} band of the Co^{4+} site ($\approx$ 0.9 eV) and e_g band of the Co^{3+} site ($\approx$ 1.9 eV) [16]. The photon energy of the pump light we used is $\approx$ 1.6 eV, suggesting that photoexcitation by the pump injects e_g carriers in the Co^{3+} site in the LS state.

Keeping those in mind, we discuss the photoexcited state for $Pr_{0.5}Ca_{0.5}CoO_3$. In Fig. 4.4b, we show the time dependence of $\Delta R/R$ at 0.5 eV measured at 30 K (LS insulating phase). Just after the photoexcitation ($\approx$ 0 ps), $\Delta R/R$ showed a sudden jump up to 20% as denoted by a closed triangle (the power density of the irradiated pump pulse was $\approx$ 6 mJ/cm^2, which corresponds to $\approx$ 0.12 photons per Co site.) From the value of the instant change of $\Delta R/R$ just after the photoirradiation, we calculated the transient reflectivity (R^{T}) with the relation that $R^{\mathrm{T}} = (1 + \Delta R/R)R(T = 30\mathrm{K})$. We conducted this ultrafast reflectance measurement from 0.5 eV to 2.1 eV, and constructed a transient spectrum of reflectivity ($R^{\mathrm{T}}(\omega)$) as shown by the closed circles in Fig. 4.4a.

For comparison, we plotted the static reflectivity at 30 and at 290 K with dashed lines. Upon the photoirradiation, the spectral shape of $R^{\mathrm{T}}(\omega)$ partly approached that of the metallic phase in the mid-infrared region, and $R^{\mathrm{T}}(\omega)$ was comparable to the metallic one above $\approx$ 1.2 eV whereas smaller below $\approx$ 1.2 eV.

To quantitatively understand the spectral shape of $R^{\mathrm{T}}(\omega)$ just after the photoirradiation, we considered the surface state generated by the photoexcitation. From $\sigma(\omega)$ at 30 K[19], the penetration depth of the pump light (d) was estimated as $\approx$ 60 nm. It is worth noting that $R^{\mathrm{T}}(\omega)$ was determined not only by the dielectric function of the photoinduced metallic phase ($\varepsilon^{\mathrm{M}}(\omega)$) but also by that of the initial insulating state

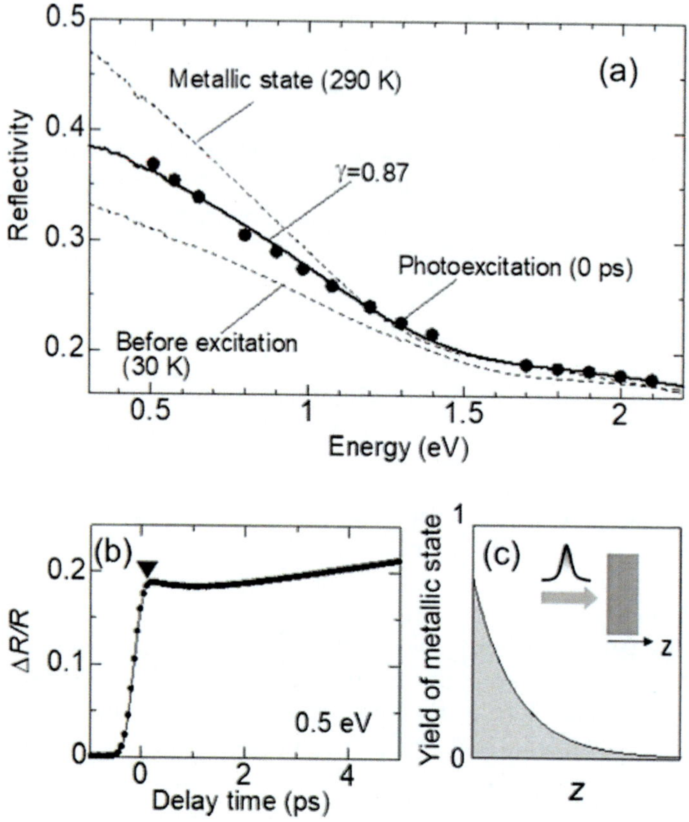

Fig. 4.4 **a** Transient reflectivity spectra just after the photoexcitation (filled circles). For comparison, linear reflectance spectra at 30 and 290 K are plotted with dashed lines. Bold line is calculated reflectivity with γ=0.87 (see text). **b** Time dependence of reflectivity change at 0.5 eV. **c** Spatial variation of the fraction of the photoinduced metallic state along z (γ=0.87). The inset shows relationship among the crystal, irradiated pulse and z axis. Reprinted with permission from [19] ©(2009) by the American Physical Society

($\varepsilon^{\mathrm{I}}(\omega)$) that partially existed even after the photoillumination (30 K). Under these circumstances, we can consider the total dielectric function of the sample, $\varepsilon(z)$ to be expressed by the following relation:

$$\varepsilon(z) = \gamma \exp(-z/d)\varepsilon^{\mathrm{M}}(\omega) + [1 - \gamma \exp(-z/d)]\varepsilon^{\mathrm{I}}(\omega) \tag{4.1}$$

$$\equiv F[\varepsilon^{\mathrm{M}}, \varepsilon^{\mathrm{I}}, \gamma\,;\, z]. \tag{4.2}$$

Here, γ is the efficiency of the photoinduced phase transition ($0 < \gamma < 1$), and z is the distance from the sample surface. The first component is created by the photoexcitation, as schematically depicted in Fig. 4.4c. We assumed that $\varepsilon^{\mathrm{M}}(\omega)$ cor-

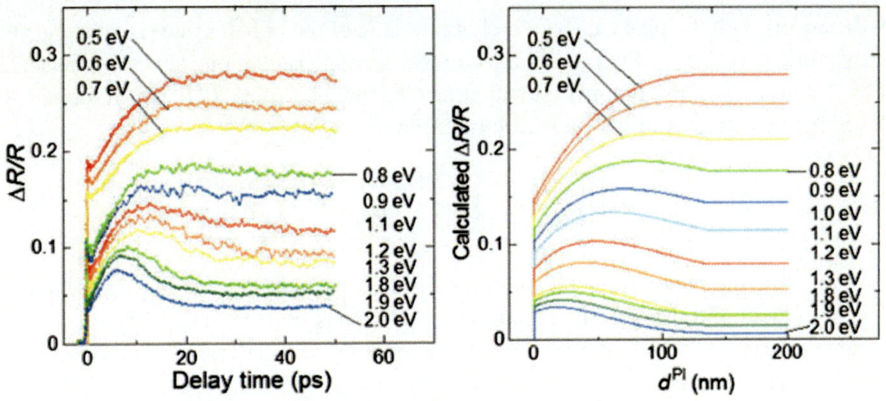

Fig. 4.5 **a** Relative change of reflectivity ($\Delta R/R$) after photoexcitation from 0.5 eV to 2.0 eV on picosecond time scale. **b** Numerically calculated results of d^{PI} dependence of $\Delta R/R$ from 0.5 eV to 2.0 eV. d^{PI} denotes thickness of uniformed metallic region. Reprinted with permission from [19] ©(2009) by the American Physical Society

responded to the dielectric function at 290 K (a metallic phase). ($\varepsilon^{\mathrm{M}}(\omega)$ and $\varepsilon^{\mathrm{I}}(\omega)$ were calculated based on Kramers-Kronig analysis from the reflectivity in Fig. 4.4a and abbreviated as ε^{M} and ε^{I} in Eq. (4.1), respectively.) This model is called a simple mixing model [20] and was used for the photoexcited states in a charge transfer complex [21] and a manganese oxide [22].

Once $\varepsilon(z)$ is determined, we can calculate reflectivity at the surface ($z = 0$) by numerically solving Maxwell's equations for the material. We calculated $R^{\mathrm{T}}(\omega)$by changing the γ values and, by least squared analysis, we found that $\gamma = 0.87$ best reproduced the observed $R^{\mathrm{T}}(\omega)$ as shown by the bold line in Fig. 4.4a. This clearly indicates the creation of a photoinduced metallic state on the fs time scale as well as the validity of the assumed $\varepsilon(z)$ expressed in Eq. (4.1).

As mentioned above, the pump light can cause a charge transfer excitation from the O $2p$ band to the empty Co e_g band and the photoexcitation can generate e_g carriers in the LS ground state and probably drive the change of the spin state and the resultant insulator-metal transition.

The next issue to be discussed is the dynamics of the metallic state after photoirradiation on the time scale of ps. In Fig. 4.5a, we show the time dependence of $\Delta R/R$ from 0 to 50 ps over a wide photon energy region, 0.5 eV–2.0 eV. At 0.5 eV, after the sudden jump, $\Delta R/R$ gradually increased and saturated at $\approx$ 20 ps. The total change in R was $\approx 27\%$ at 50 ps. With the increasing photon energy of the probe light, however, the magnitude of $\Delta R/R$ gradually decreased, and, more importantly, formed a broad peak and decreased with further delay time. The peak time in the time profile decreased with the photon energy to, for example, $\approx$ 6.5 ps at 2.1 eV. Above $\approx$ 30 ps, for all the photon energies, $\Delta R/R$ barely changed with the delay time, indicating saturation of the photonic change. The observed time profile with the conspicuous dependence on the probe energy is characteristic of the cobalt system and suggests a dynamic variation of the photoinduced metallic state on the picosecond time scale.

To describe this unconventional time dependence of $\Delta R/R$, we assumed the following variation of the dielectric function.

$$\varepsilon(z) = \begin{cases} F[\varepsilon^{\mathrm{M}}, \varepsilon^{\mathrm{I}}, \gamma\,;\ 0] & (0 < z < d^{\mathrm{PI}}) \\ F[\varepsilon^{\mathrm{M}}, \varepsilon^{\mathrm{I}}, \gamma\,;\ z - d^{\mathrm{PI}}] & (d^{\mathrm{PI}} < z). \end{cases} \tag{4.3}$$

Here, d^{PI} is the thickness of the photonically created metallic state. Again, by numerically calculating Maxwell's equations on the basis of $\varepsilon(z)$ defined in Eq. (3), we calculated $R(\omega)$ at the surface ($z = 0$) as a function of d^{PI}. (The γ value was fixed at 0.7 to reproduce the observed time dependence.) In Fig. 4.5b, we show the calculated $R(\omega)$ as a function of d^{PI} up to $d^{\mathrm{PI}} = d^{\mathrm{fin}}$ ($\approx$ 130 nm) in the photon energy region 0.5 eV –2 eV. (We extrapolated a constant value above d^{fin}.) As can clearly be seen, the numerical calculation agrees well with the above-mentioned features observed in the time profiles of $\Delta R/R$ [see Fig. 4.5a], assuming that d^{PI} corresponds to the delay time. With an increase in d^{PI}, the magnitude of $\Delta R/R$ gradually decreased and showed a broad peak, as denoted by the blue triangles, and the peak moved to the left. The clear similarity between the plots in Fig. 4.5a, b strongly indicates that d^{PI} can be identified with the delay time, meaning that the photonically excited metallic state expands along z.

The schematics in Fig. 4.6a–c illustrate the evolution of the photoinduced metallic region as well as the assumed $\varepsilon(z)$. Just after the photoirradiation, the metallic state (yellow region) occurred around the surface of the material (dark brown region) as shown in Fig. 4.6a. As time elapsed [Fig. 4.6b], the metallic region showed self-propagation along z, indicating the movement of the domain wall characterized by d^{PI} between the metallic and original insulating state. The peaks observed in the $\Delta R/R$ profile resulted from the interference of the probe light within the photoinduced metallic region. The propagation saturated around $d^{\mathrm{PI}} \approx 130$ nm as seen in Fig. 4.6c.

To estimate the time scale of the development of the photoinduced metallic region, we plotted the relationship between the delay time and the d^{PI} at which $\Delta R/R$ shows a peak in Fig. 4.5a, b at all photon energies. As shown in Fig. 4.6d, d^{PI} approximately linearly increased with d^{PI}. The velocity was estimated, using least squares fitting analysis, to be $\approx 4.4 \times 10^3$ m/s. This value is comparable to the sound velocity estimated from the elastic constant of a similar cobalt perovskite [23]. It is worth noting here that the propagation speed is faster than that of the expansion of the thermally equilibrium state because the heat conduction successively occurred after the phonon propagation and the speed of the heated region was even slower than the above sound velocity. The striking correspondence with the sound velocity implies that the local expansion of the lattice generated by the photoillumination drove the real space propagation of the photoinduced metallic region.

Note that even after the damping of the acoustic phonon, the photoinduced metallic region still survived because of the electronic stability of the intermediate metallic state. In this sense, the acoustic phonon in the present system behaved as if it were applying negative pressure, which can only be achieved coherently by ultrafast

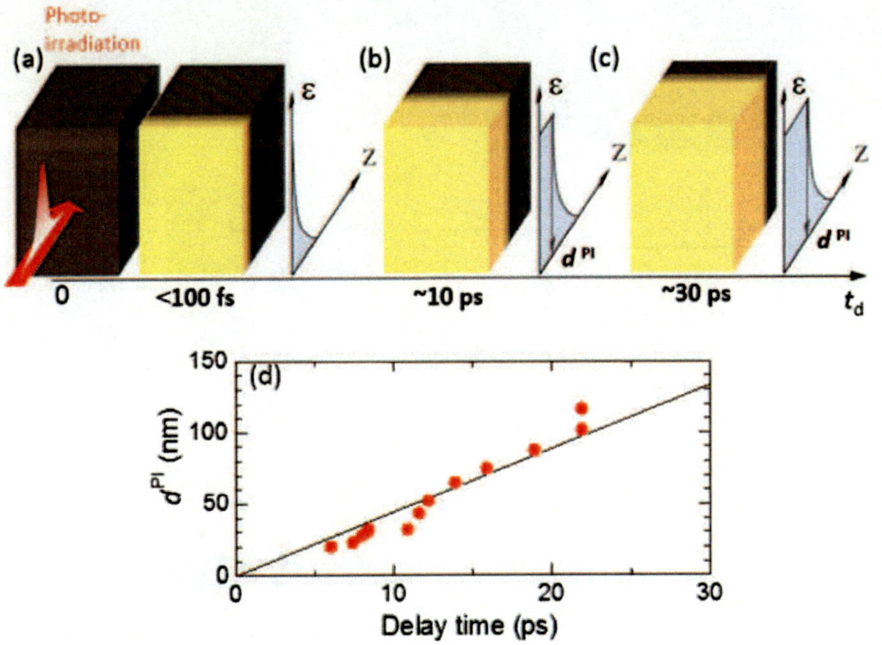

Fig. 4.6 **a–c** Schematics of time evolution of photonically created metallic region, together with dielectric function ($\varepsilon(z)$). Yellow regions show photoinduced metallic region that propagates itself along z with time evolution. **d** Correlation between delay time and d^{PI}. Solid line shows the result of least squares fitting analysis. Reprinted with permission from Ref. [19] ©(2009) by the American Physical Society

excitation of acoustic phonon wave packets. Accordingly, $Pr_{0.5}Ca_{0.5}CoO_3$ can be classified as a unique example of a real space photoinduced phase transition, which may provide clues to understanding various ultrafast non-equilibrium phenomena.

4.2.4 Real Space Control of the Photoexcited State

Up to the previous sections, we described how the photoinduced metal phase of $Pr_{0.5}Ca_{0.5}CoO_3$ was generated and propagated in a real space. From a different point of view, this can be viewed as a photoinduced fabrication of a metallic film on $Pr_{0.5}Ca_{0.5}CoO_3$ substrate and an observation of the dynamics of the formation. Furthermore, it is also possible to make multi-layered photoinduced thin film by multistage excitation, in which the film thickness can be controlled by the optical excitation interval. In this section, we describe the study of double excitations for $Pr_{0.5}Ca_{0.5}CoO_3$ using successive fs pulses and demonstrate a photonic heterostructure composed of photoexcited states. It is noteworthy that such a photonic thin film is not chemically but rather electronically created, indicating that it can be rewritable

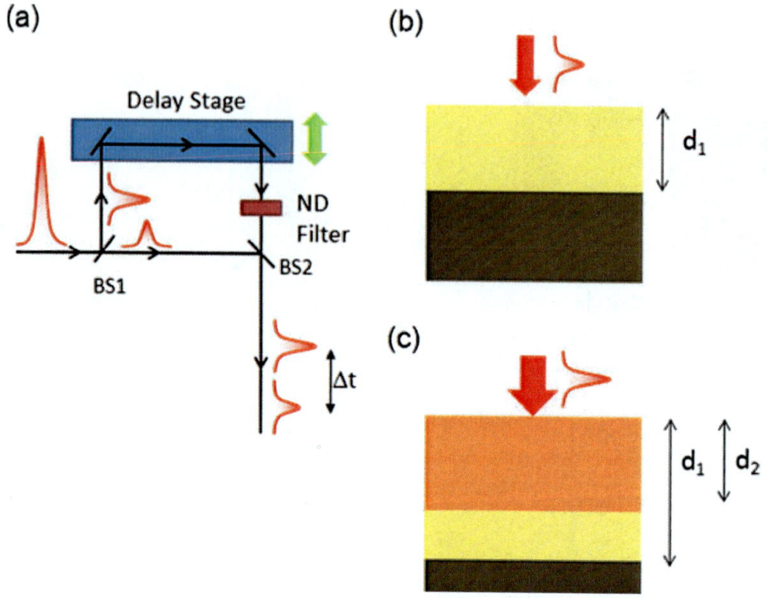

Fig. 4.7 **a** Schematic diagram for optics to create sequential pulses. Δt denotes interval of sequential pulses. **b**, **c** Schematics of sample after first (**b**) and second (**c**) excitation. d_1 and d_2 denote thickness of each propagated domain

and that the electronic nature can be minutely controlled with well-tuned fs laser pulses. This is promising in terms of ultrafast optical communication devices.

From this viewpoint, we prepared two successive fs pulses as shown in Fig. 4.7a for the excitation. As a light source, we used a mode-locked Ti:sapphire laser pulse ($\approx$ 1.6 eV). The amplified single pulse was divided by a beam splitter into two pulses. One pulse was converted into 2.0 eV by OPA to monitor $\Delta R/R$ after the pumping. The other was used as a pump light for the excitation. For the sequential excitation, we further divided the pump pulse with a first beam splitter (BS1) as seen in Fig. 4.7a. The intervals between the divided pulses were controlled by a delay stage, and the two pulses were finally made into sequential pulses in a uniaxial line through a second beam splitter (BS2). Using the successive pulses, we excited the sample sequentially.

For probing the doubly photoexcited state, we again performed the time dependence of $\Delta R/R$ with the pump-probe technique like the experiment described in the previous section. $\Delta R/R$ was monitored with the 2.0 eV pulse that was firstly divided and converted by OPA. All the measurements were done at 30 K.

Figures 4.8a–c show the time dependence of $\Delta R/R$ after the sequential pulse excitations. In all the profiles, the sample surface was irradiated with the first pulse at ≈ 0 ps as indicated by the arrows on the horizontal axes. Each fluence of the first excitation was $\approx$ 2 mJ/cm^2 (a), $\approx$ 3 mJ/cm^2 (b), and $\approx$ 4 mJ/cm^2 (c). After that, we irradiated the surface again with a second pulse, whose fluence was fixed at a

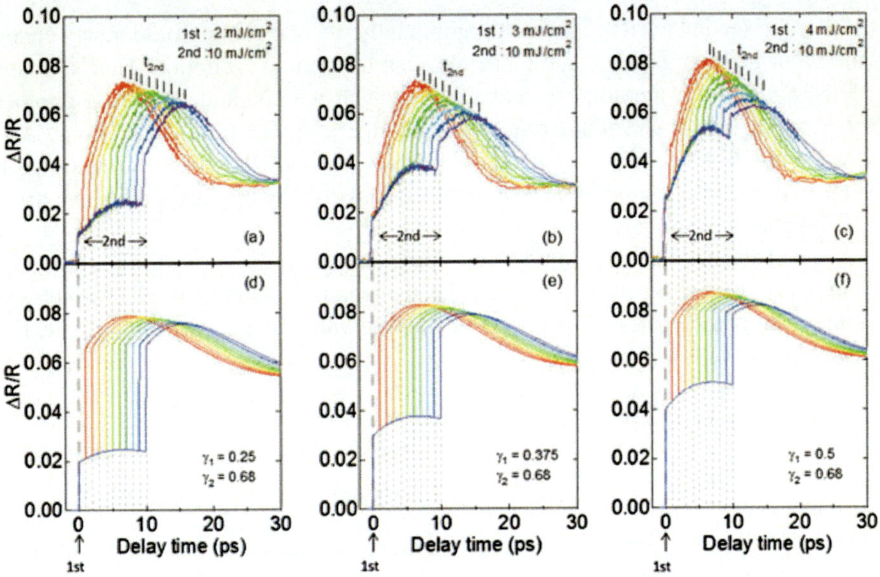

Fig. 4.8 **a–c** Time profiles of $\Delta R/R$ after sequential excitations. Fluences of first excitation were ≈ 2 mJ/cm^2 (**a**), ≈ 3 mJ/cm^2 (**b**), and ≈ 4 mJ/cm^2 (**c**), and second was fixed at ≈ 10 mJ/cm^2. Vertical dashed lines denote timings of first and second excitations. **d–f** Calculated time profiles of $\Delta R/R$ with two parameters, γ_1 and γ_2, denoting efficiencies of photonic transition by first and second excitation, respectively. Values of γ_1 are 0.25 (**d**), 0.375 (**e**), and 0.5 (**f**), and γ_2 is fixed at 0.68

constant value (≈ 10 mJ/cm^2), several intervals from the first pulse (Δt). At the first excitation, the $\Delta R/R$ profiles show an instant jump on account of the photonic formation of the metallic domain as shown in the previous subsection. Δt ps after the first photoexcitation, the sample was irradiated again with the second pulse as denoted by vertical dashed lines. We observed a sharp jump similar to the first excitation and successive formation of a broad peak in $\Delta R/R$ at t_{2nd}, as shown by the vertical bars in Fig. 4.8a–c. It is reasonable to assume that the broad peak was caused by the effect of interference within the photolayer formed by the photoexcitations, but the interval between the time of the second excitation and $t_{2\mathrm{nd}}$ became shorter than in the case of excitation using a single pulse.

To understand the profiles of $\Delta R/R$ after doubled pulse excitations, let us consider the spatial dependence of the total dielectric function $\varepsilon(z)$ again, z being the distance from the surface. In a single pulse excitation, the time profile of $\Delta R/R$ can be well described in terms of Eq. (4.3),

$$\varepsilon(z) = \begin{cases} F[\varepsilon^{\mathrm{M}}, \varepsilon^{\mathrm{I}}, \gamma_1 ;\ 0] & (0 < z < d_1) \\ F[\varepsilon^{\mathrm{M}}, \varepsilon^{\mathrm{I}}, \gamma_1 ;\ z - d_1] & (d_1 < z), \end{cases} \tag{4.4}$$

where d_1 and γ_1 are the thickness of the propagated metallic state and efficiency of the photonic change by the first excitation [see Fig. 4.7b].

For the second excitation, we thought that the second pulse should newly change the LS insulating portion that remained even after the first excitation. This is because the $\Delta R/R$ in the metallic state at 290 K, which we concluded to be a photoexcited state, hardly showed a reflectance change after the photoirradiations. Also, we assumed independent propagation after the second excitation as illustrated by Fig. 4.7c because the domain movement was assisted by an ultrasonic wave driven by the photoexcitation.

The efficiency of the photoinduced phase change in the second excitation is γ_2, and hence, total efficiency (γ_{all}) after the sequential excitations is expressed by the relationship $\gamma_{all} = \gamma_1 + (1 - \gamma_1)\gamma_2$). Using the value of γ_{all}, the total $\varepsilon(z)$ can be described as follows.

$$\varepsilon(z) = \begin{cases} F[\varepsilon^{\mathrm{M}}, \varepsilon^{\mathrm{I}}, \gamma_{all}\,;\ 0] & (0 < z < d_2) \\ F[\varepsilon^{\mathrm{M}}, \varepsilon', \gamma_2\,;\ z - d_2] & (d_2 < z < d_1, \quad \varepsilon' = F[\varepsilon^{\mathrm{M}}, \varepsilon^{\mathrm{I}}, \gamma_1\,;\ 0]) \\ F[\varepsilon^{\mathrm{M}}, \varepsilon^{\mathrm{I}}, \gamma_1\,;\ z - d_1] & (d_1 < z_1). \end{cases} \quad (4.5)$$

Once $\varepsilon(z)$ is determined, $\Delta R/R$ can be numerically calculated as a function of d_1 and d_2, which can be converted into the delay time by using the domain velocities for each respective fluence. In Figs. 4.8d–f, we show the calculated time profiles of $\Delta R/R$. To reproduce the experimental results, the values of γ_1, which corresponds to the fluence at the first excitation, were set at 0.25 (d), 0.375 (e), and 0.5 (f), and γ_2, the fluence at the first excitation, was fixed at 0.68. The calculated profiles qualitatively reproduced the observed $\Delta R/R$. To see the doubly excited profiles more quantitatively, we compared the calculation with the experiment in terms of the peak time, $t_{2\mathrm{nd}}$, observed for the $\Delta R/R$ profile after the second excitation as shown by the vertical lines in Fig. 4.8a–c. In Fig. 4.9a, we plotted the values of $t_{2\mathrm{nd}}$ as a function of Δt, the interval of the two pump pulses, as closed circles. The value of $t_{2\mathrm{nd}}$ gradually and concavely increased as Δt increased. For comparison, we also plotted the peak time without the first excitation [i.e., 6.5 +Δt (ps), 6.5 ps being the peak time with a single excitation with the fluence of $\approx$ 10 mJ/cm^2] represented by the solid line. If the first excitation did not exist, the filled circles should coincide with the solid line. In the experiment, the value of $t_{2\mathrm{nd}}$ was smaller than that in the case of single excitation in the whole region of Δt and further decreased as the fluence increased.

Figure 4.9b shows the calculated Δt dependence of $t_{2\mathrm{nd}}$ estimated from Fig. 4.8d–f, together with the peak time without the contribution of γ_1, corresponding to the first excitation. The calculation well reproduced the three experimentally observed features concerning $t_{2\mathrm{nd}}$: the concave increase, the suppression by the sequential excitations, and the further decrease as the fluence increases. These results are due to the interference effect of the probe light that occurred among the accumulated layers and affected the shape of the time profiles of $\Delta R/R$ and hence the value of $t_{2\mathrm{nd}}$. These results show the formation of the photonic superlattice as modeled in Eq. (4.5), indicating that the dielectric function or refractive index in a layer can be controlled by changing the fluence of excitation light, and that the spatial pattern of the photoexcited state can be freely designed on the scale of the wavelength of pump light. This enables us to fabricate various dynamical and rewritable photonic

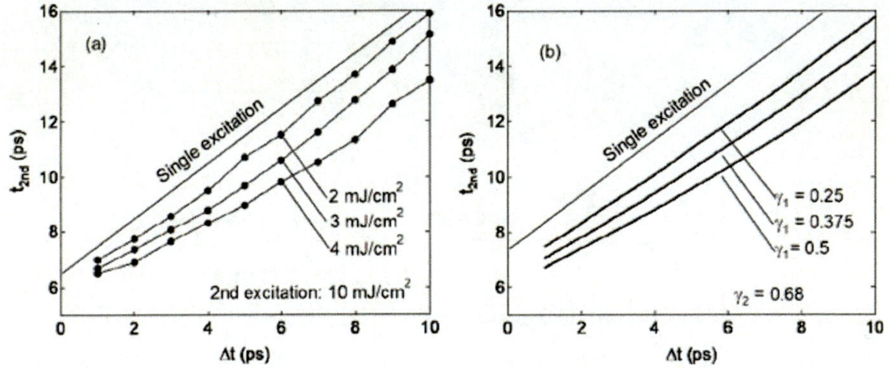

Fig. 4.9 (Δt dependence of peak time after second excitation (t_{2nd}) estimated from experimental (**a**) and calculated (**b**) $\Delta R/R$ (closed circles) at selected fluence. Solid lines denote peak time $\Delta R/R$ after second excitation ($\approx$ 10 mJ/cm) without first excitation (see text)

crystals or optical waveguides on $Pr_{0.5}Ca_{0.5}CoO_3$, which can be viewed as one of the applications of the photoinduced phenomena to possible ultrafast communication devices.

4.3 Photonic Control of Polar Cobalt Perovskite $BiCoO_3$

4.3.1 Introduction

Solid materials with a polar structure have attracted much interest in condensed matter physics because polar materials including ferroelectrics show various interesting mechanical, electrical, and optical phenomena originating from the breaking of the inversion symmetry. Some SC cobalt perovskites also show a polar structure and ferroelectric behaviors, and $BiCoO_3$ is one of the examples described in Chap. 6 of this book. Recently, Azuma and his coworkers reported a new class of SC cobaltite, $BiCoO_3$ [24]. Figure 4.10a shows the crystal structure of $BiCoO_3$ at room temperature. The crystal symmetry is P4mm, which is the same symmetry as the famous ferroelectrics lead titanates, $PbTiO_3$ or PZT. Compared with a typical perovskite, one of the apical oxygen ions is shifted in the direction of the c-axis [see Fig. 4.10a], and a CoO_5 pyramid exists in the unit cell. This clearly suggests that $BiCoO_3$ has a polar structure along the c-axis, which realizes gigantic spontaneous polarization at about $\approx 120\mu$C/cm at room temperature.

In $BiCoO_3$, both the Bi and Co ions are trivalent, and the spin configuration of Co^{3+} is in the HS state due to the pyramidal structure, where the degenerated e_g and t_{2g} states are lifted as shown in Fig. 4.10b [24]. An important aspect of $BiCoO_3$ is the pressure effect; while temperature change hardly affects the spin state, applying external pressure on $BiCoO_3$ remarkably changes the spin configuration in Co^{3+}

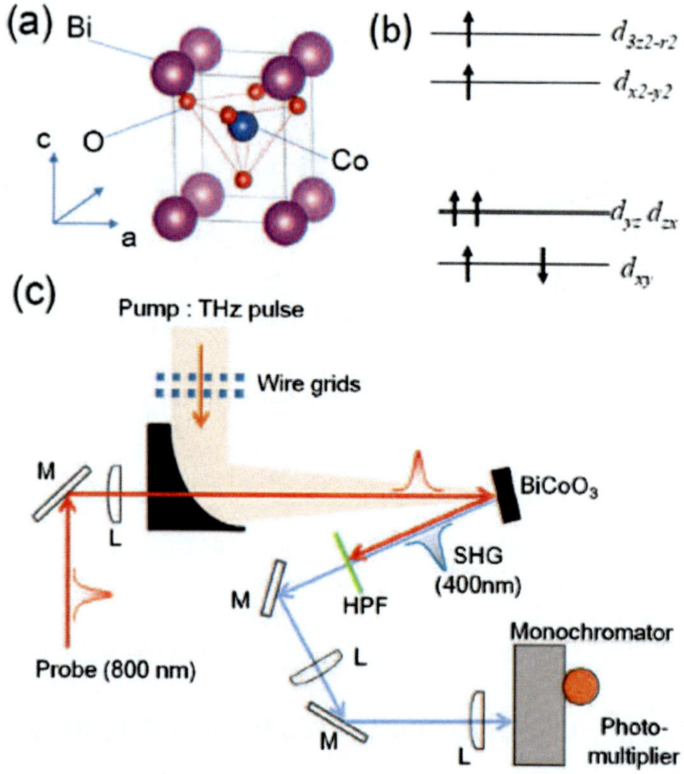

Fig. 4.10 Crystal structure (**a**) and spin configuration (**b**) of $BiCoO_3$. **c** Schematics of THz pump and SHG probe measurement system. THz pump pulse was focused onto the sample by parabolic mirror during 800 nm probe pulse with lens through parabolic mirror. Generated SH pulse was introduced to a monochromator and detected by photomultiplier. Reprinted with permission from Ref. [32] ©(2017) by the American Physical Society

from the HS state to a lesser spin state (the IS or LS state) with lattice contraction [24]. Oka *et al*. also revealed that the pressure-induced SC change was involved in the polar to nonpolar structural change, indicating that $BiCoO_3$ not only belongs to the SC family but also a new class of strongly correlated polar material, which is fully described in Chap. 6.

As is widely known, polar material shows second-order nonlinear optical phenomena, such as second harmonic generation (SHG), a photovoltaic effect, the generation of THz pulses, etc. due to finite second-order electric susceptibility $\chi^{(2)}$ originating from the polar structure. Thanks to the recent development of fs laser technology, we can see the nonlinear effects more easily in many polar materials.

In this section, we show ultrafast control of the SHG signal in $BiCoO_3$ by irradiation of fs terahertz (THz) laser pulses as an external perturbation. As for the THz pulse, the recent development of this ultrafast laser technique [25] can generate strong electromagnetic waves whose electric field reaches up to an order of ≈ 1 MV/cm in the THz photon energy region. Owing to such a strong THz field, some intriguing phenomena have recently been reported, e.g., the dynamical Frantz–Keldysh effect in semiconductors [26, 27], the observation of the Higgs mode [28], insulator-metal phenomena in a correlated oxide [29, 30], and photoinduced superconductivity [31]. Here, with an ultrafast laser technique and novel polar cobaltite, we demonstrate gigantic and ultrafast variations of the SHG signal in $BiCoO_3$ with THz pulses [32] as another example of applying THz pulses.

4.3.2 *Experiment*

A polycrystalline sample of $BiCoO_3$ was fabricated under high pressure of $\approx$ 6 GPa and at 1200°C, and the details are provided in Chap. 6 as well as in Ref. [24].

The relative changes in reflectivity, $\Delta R/R$, and the SHG intensity ($\Delta I_{SH}/I_{SH}$) after the optical pump of THz pulse were obtained with the pump-probe method using a mode-locked Ti:sapphire laser light equipped with a regenerative amplifier (pulse width of $\approx$ 150 fs, repetition rate of 1 kHz, and photon energy of $\approx$ 1.55 eV). Figure 4.10c shows schematics of a system for measuring the pump-probe SHG change. The amplified pulse was separated into two light beams; one was used to generate the pump THz pulse (see below). The other beam was used as a probe pulse. We irradiated $BiCoO_3$ with the probe pulse through the parabolic mirror and detected the reflectance and SHG change in accordance with the THz pumping. For the reflectance measurement, the reflected light was detected by an Si photodiode, and for the SHG measurement, the frequency-doubled light ($\approx$ 3.1 eV) was detected by a photomultiplier. The fundamental $\approx$ 1.55 eV pulses were deleted by high-pass filters and a grating-type monochromator set in the optical path.

For the photoexcitation, we generated THz pulses with the tilted pulse front method with a $LiNbO_3$ crystal and the separated amplified pulse. The details on the procedure for generating the THz pulses are provided in Ref. [25]. The time profile of the THz pulse was detected by the electro-optical sampling method. The generated THz pulses were focused onto the sample with a parabolic mirror as shown in Fig. 4.10c. All the experiments were done in air and at room temperature.

4.3.3 *Optical Control of the Polarity with Terahertz Pulse Excitation*

In Fig. 4.11b, we show temporal profiles of the applied THz field (E_{THz}) that was used for the optical excitation. For the orange, green, and blue profiles, the intensity of each field was 75%, 50%, and 25% of that for the red profile, respectively. (Some structures seen after 1 ps in the profile were caused by water in air.) For the red

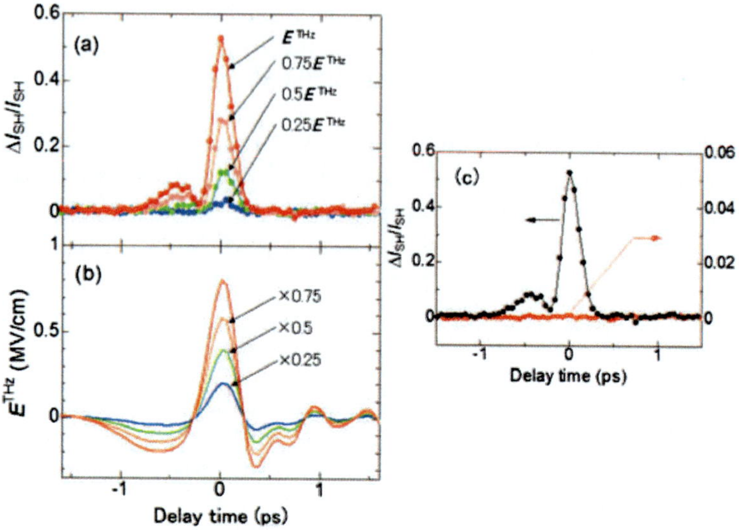

Fig. 4.11 **a** Time profiles of relative change of second harmonic intensity ($\Delta I_{SH}/I_{SH}$) with several different THz pulses. Color of each profile corresponds to that of THz pulse shown in (**b**). **b** Waveforms of THz field used for photoexcitation in time domain. For orange, green, and blue profiles, intensities of field were 3/4, 1/2, and 1/4 that of red profile, respectively. **c** Time profile of relative change of reflectivity ($\Delta R/R$, red circles), together with that of $\Delta I_{SH}/I_{SH}$ (black circles). Reprinted with permission from Ref. [32] ©(2017) by the American Physical Society

profile, the maximum value of the THz electric field (E^{peak}) reached up to ≈ 0.8 MV/cm at ≈ 0 ps.

Figure 4.11a shows the time evolution of $\Delta I_{SH}/I_{SH}$ in $BiCoO_3$ with the irradiation of the THz pulse shown in Fig. 4.11b. Applying E_{THz}, $\Delta I_{SH}/I_{SH}$ showed a remarkable increase, and the magnitude of the relative change reached more than 50% at ≈ 0 ps as shown by the red profiles. As the E_{THz} decreased, $\Delta I_{SH}/I_{SH}$ decreased nonlinearly. What should be noticed is the sign of $\Delta I_{SH}/I_{SH}$. In $BiCoO_3$, $\Delta I_{SH}/I_{SH}$ is always positive, regardless of the sign of the THz pulse.

This contrasts with the results obtained for some organic ferroelectrics [33, 34]. For comparison, in Fig. 4.12a, b we show time profiles of SHG change in Hdppz-Hca cocrystal together with THz waveform applied [34]. Hdppz-Hca is a novel ferroelectric cocrystal ($T_c \approx 402$K) composed of protonated 2,3-di(2-pyridinyl)pyrazine (Hdppz) and deprotonated chloranilic acid (Hca) [35]. With applying THz field of ≈ 1 MV/cm, the sign of $\Delta I_{SH}/I_{SH}$ exactly followed the direction of the applied THz field. This indicates that the THz pulse in Hdppz-Hca directly modulates the degree of the polarization. Similar linear dependence on THz field is observed in other organic ferroelectics, TTF-CA [33]. Those observations are in contrast to the results seen in $BiCoO_3$.

In Fig. 4.11c, we show the time profile of $\Delta R/R$ with the irradiation of a THz pulse shown by the red curve in Fig. 4.11b, together with the profile of $\Delta I_{SH}/I_{SH}$. While the magnitude of $\Delta I_{SH}/I_{SH}$ was over 50%, $\Delta R/R$ scarcely varied ($< 0.1\%$)

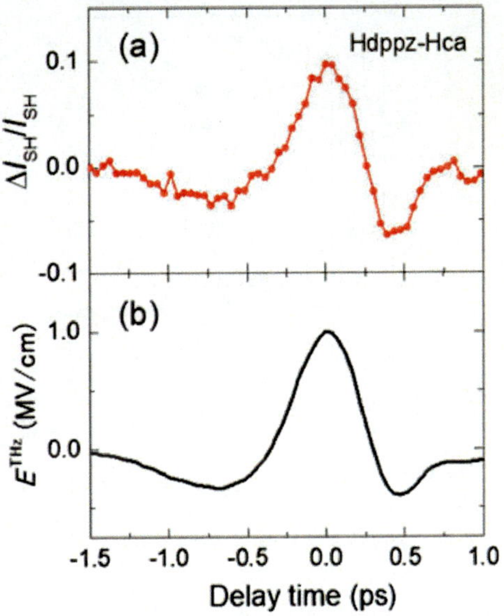

Fig. 4.12 **a** Time profiles of the relative change of second harmonic intensity ($\Delta I_{SH}/I_{SH}$) with the THz field displayed in (**b**) in Hdppz-Hca cocrystal. **b** Waveforms of THz field used for photoexcitation in time domain. The data were taken from Ref. [34]

with the same THz field. This signals that the applied field hardly changed the linear refractive index n or absorption coefficient (α) at 3.1 eV and, hence, the change of $\Delta I_{SH}/I_{SH}$ could not have originated from the simple photonic change of n and/or α, that is, the first order of electric susceptibility, $\chi^{(1)}$.

In Fig. 4.13b, to see the variation of the SHG change more quantitatively, we plotted the normalized maximum value (ΔI_{SH}^{peak}) as a function of the peak electric field at ≈ 0 ps, E^{peak}. The value of ΔI_{SH}^{peak} nonlinearly increased with increase in E^{peak}. The E^{peak} dependence of the ΔI_{SH}^{peak} was well fit by the parabolic curve as shown by the solid line, indicating that $\Delta I_{SH}^{peak} \propto E^{peak}$.

Figure 4.13a demonstrates the time profile of $\Delta I_{SH}/I_{SH}$ (black circles) together with that of $|E_{THz}|^2$ (red line). $\Delta I_{SH}/I_{SH}$ was well scaled by $|E_{THz}|^2$. These results also signal that the observed SH change was proportional to $|E_{THz}|^2$ for all the delay time.

Here, let us discuss the origin of the observed THz field effect. In contrast with the above-mentioned organic ferroelectrics [33, 34], $BiCoO_3$ always shows a positive increase in I_{SH} regardless of the direction of E_{THz}, implying that the THz pulse affects $\chi^{(2)}$ itself through the optical transition realized by the THz pulse. Judging from the HS configuration in $BiCoO_3$ [Fig. 4.10b], it is reasonable to consider that the d–d excitation from the occupied d_{xy} to unoccupied d_{yz} or d_{zx} state can exist around the wide energy band of the applied THz pulse. (In a similar perovskite-type cobaltite $LaCoO_3$, Iwai et al. [36] measured absorption spectrum at around THz energy region and observed d–d transition in Co^{3+}.) Under these circumstances, the photoexcited d_{yz} or d_{zx} electron can shift the position of the apical oxygen ions in the pyramid through electron-phonon coupling and bring about resultant elongation of the polar pyramidal structure or increase of $\chi^{(2)}$.

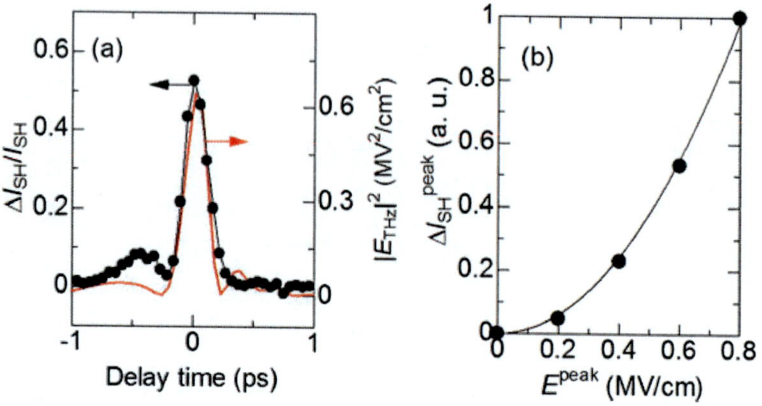

Fig. 4.13 **a** Time profiles of $\Delta I_{SH}/I_{SH}$ and square of applied THz field ($|E_{THz}|^2$ [32]. **b** Normalized maximum value of second harmonic change at ≈ 0 ps (ΔI_{SH}^{peak}) as function of peak value in THz electric field (E^{peak}) Reprinted with permission from Ref.[32] ©(2017) by the American Physical Society

When we consider that the lowest term of E_{THz}, $\chi^{(2)}$ can be expanded as $\chi^{(2)} \approx \chi_0^{(2)} + \Delta\chi^{(2)}|E_{THz}|^2$, where $\chi_0^{(2)}$ and $\Delta\chi^{(2)}$ are the original susceptibility and variation of $\chi^{(2)}$ with the THz pulse, respectively. Then, $\Delta I_{SH}/I_{SH} \approx 2|\Delta\chi^{(2)}/\chi_0^{(2)}||E_{THz}|^2$, which is consistent with the observed power dependence in Fig. 4.11a.

It is noteworthy that $\Delta I_{SH}/I_{SH}$ in Fig. 4.11a instantly decreased the moment that the applied THz field disappeared. Very recently, Morimoto and Nagaosa calculated $\chi^{(2)}$ in ferroelectrics [37] considering the Floquet state as well as Berry phase connections between the valence and conduction bands. According to their scenario, $\chi^{(2)}$ can be increased when the incident pulse forms photon-dressed excitons in a ferroelectric crystal. This idea can be applicable to our experiment with $BiCoO_3$; the interorbital transition between t_{2g} states driven by the strong THz pulse caused the photon-coupled excitonic state (THz-dressed state), which can explain not only the resultant enhancement of I_{SH} but also the observed ultrafast variation of I_{SH}. Whether lattice deformation is involved in the THz-dressed state is an interesting problem, and to confirm the scenario, for example, ultrafast structural measurements using fs X-ray pulses for probe and THz pulses for excitation would be indispensable and deserve a future study.

4.4 Summary

In this chapter, we introduced examples of the photonic control of the electronic state for two SC cobaltites, $Pr_{0.5}Ca_{0.5}CoO_3$ and $BiCoO_3$.

In $Pr_{0.5}Ca_{0.5}CoO_3$, with the excitation of charge transfer excitation ($O2p \rightarrow$ Co $3d$), we observed a large and ultrafast change in the reflectivity spectrum $R(\omega)$ just after the photoexcitation. The observed transient $R(\omega)$ and the dynamics were well reproduced on the basis of the linear combination of the dielectric function of the initial and high-temperature metallic states, indicating ultrafast photoinduced phase transition and propagation of the photonically created metallic domain on the ps time scale. In addition, we tried to make a photonic superlattice on $Pr_{0.5}Ca_{0.5}CoO_3$ by sequential pulse excitation. The multi-step time profile of $\Delta R/R$ as well as the peak position after the pumping could quantitatively be reproduced in terms of a model of the successive propagation of a photodomain, indicating fabrication of a photonic superlattice. The results observed for $Pr_{0.5}Ca_{0.5}CoO_3$ can be viewed as a novel example of real space photoinduced phase transition and, as well as the analytical method, can provide clues to understanding the various ultrafast non-equilibrium phenomena.

In a polar cobalt perovskite, $BiCoO_3$, we investigated the variation in the SHG signal through irradiation with THz pulses. The irradiation (≈ 0.8 MV/cm) caused a large increase in the SHG intensity by more than 50% at room temperature. In particular, the ultrafast enhancement of the SHG signal might be understood in terms of the photonically dressed orbital transitions between Co t_{2g} states driven by THz pulses as recent theoretical works have suggested. These results indicate ultrafast switching of the polar structure in $BiCoO_3$ as well as a method of controlling the figure of merit of nonlinear crystals.

Acknowledgements The authors appreciate R. Fukaya and K. Onda for their technical assistance throughout this work and thank M. Itoh, T. Kyomen, X. Peng, M. Tamura, K. Seko, M. Kurashima, S. Naruse, N. Yamaya, T. Umanodan, H. Hirori, K. Tanaka, K. Oka, and M. Azuma, for discussions. This work was supported by JSPS KAKENHI.

References

1. As comprehensive reviews, "Relaxations of Excited States and Photoinduced Structural Phase Transitions", editd by K. Nasu (Springer-Verlag, Berlin 1997), "Photo induced Phase Transitions", edited by K. Nasu (World Scientific publishing, Singapore, 2004) and K. Yonemitsu and K. Nasu, J. Phys. **465**, 1 (2008)
2. M. Imada, A. Fujimori, Y. Tokura, Rev. Mod. Phys. **70**, 1039 (1998)
3. K. Miyano, T. Tanaka, Y. Tomioka, Y. Tokura, Phys. Rev. Lett. **78**, 4257 (1997)
4. A. Cavalleri, M. Rini, H.H.W. Chong, S. Fourmaux, T.E. Glover, P.A. Heimann, J.C. Kieffer, R.W. Schoenlein, Phys. Rev. Lett. **95**, 067405 (2005)
5. For example, *"Spin Crossover in Transition Metal Compounds I, II and III", edited by P* (Goodwin (Springer Verlag, Berlin Heidelberg, Gutlich and H. A, 2004)
6. K. Asai, O. Yokokura, N. Nishimori, H. Chou, J.M. Tranquada, G. Shirane, S. Higuchi, Y. Okajima, K. Kohn, Phys. Rev. B **50**, 3025 (1994)
7. S. Yamaguchi, Y. Okimoto, H. Taniguchi, Y. Tokura, Phys. Rev. B **53**, R2926 (1996)
8. Saitoh, T. Mizokawa, A. Fujimori, M. Abbate, Y. Takeda, and M. Takano: Phys. Rev. B **55** (1997) 4257;
9. A. Slimani, F. Varret, K. Boukheddaden, C. Chong, H. Mishra, J. Haasnoot, and S. Pillet Phys. Rev. B **84**, 094442 (2011)

10. R. Bertoni, M. Lorenc, H. Cailleau, A. Tissot, J. Laisney, M.L. Boillot, L. Stoleriu, L. Stancu, C. Enachescu, E. Collet, Nat. Mat. **15**, 606 (2016)
11. M.A. Korotin, S.Y. Ezhov, I.V. Solovyev, V.I. Anisimov, D.I. Khomskii, G.A. Sawatzky, Intermediate-spin state and properties of $LaCoO_3$. Phys. Rev. B **54**, 5309 (1996)
12. S. Yamaguchi, Y. Okimoto, Y. Tokura, Phys. Rev. B **55**, R8666 (1997)
13. S. Tsubouchi, T. Kyomen, M. Itoh, P. Ganguly, M. Oguni, Y. Shimojo, Y. Morii, and and Y. Ishii, Phys. Rev. B **66**, 052418 (2002)
14. S. Tsubouchi, T. Kyomen, M. Itoh, M. Oguni, Phys. Rev. B **69**, 144406 (2004)
15. T. Fujita, T. Miyashita, Y. Yasui, Y. Kobayashi, M. Sato, E. Nishibori, M. Sakata, Y. Shimojo, N. Igawa, Y. Ishii, K. Kakurai, T. Adachi, Y. Ohishi, M. Takata, J. Phys. Soc. Jpn. **73**, 1987 (2004)
16. T. Saitoh, Y. Yamashita, N. Todoroki, T. Kyomen, M. Itoh, M. Higashiguchi, M. Nakatake, K. Shimada, J. El. Spectrosc. **144**, 893 (2005)
17. J. Hejtmanek, Z. Jirak, O. Kaman, K. Knızek, E. Santava, K. Nitta, T. Naito, H. Fujishiro, Eur. Phys. J. B **86**, 305 (2013)
18. J. Kunes, P. Augustinsky, Phys. Rev. B **90**, 235112 (2014)
19. Y. Okimoto, X. Peng, M. Tamura, T. Morita, K. Onda, T. Ishikawa, S. Koshihara, N. Todoroki, T. Kyomen, M. Itoh, Phys. Rev. Lett. **103**, 027402 (2009)
20. For example, H.Ch. Weissker, J. Furthmuller, and F. Bechstedt, Phys. Rev. B **67**, 165322 (2003)
21. H. Okamoto, T. Miyagoe, K. Kobayashi, H. Uemura, H. Nishioka, H. Matsuzaki, A. Sawa, Y. Tokura, Phys. Rev. B **83**, 125102 (2011)
22. Y. Okimoto, H. Matsuzaki, Y. Tomioka, I. Kezsmarki, T. Ogasawara, M. Matsubara, H. Okamoto, Y. Tokura, J. Phys. Soc. Jpn. **76**, 043702 (2007)
23. S. Murata, S. Isida, M. Suzuki, Y. Kobayashi, K. Asai, K. Kohn, Physica B **263–264**, 647 (1999)
24. K. Oka, M. Azuma, W. Chen, H. Yusa, A.A. Belik, E.T. Muromachi, M. Mizumaki, N. Ishimatsu, N. Hiraoka, M. Tsujimoto, M.G. Tucker, J.P. Attfield, Y. Shimakawa, J. Am. Chem. Soc. **132**, 9438 (2010)
25. H. Hirori, A. Doi, F. Blanchard, K. Tanaka, Appl. Phys. Lett. **98**, 091106 (2011)
26. H. Hirori, M. Nagai, K. Tanaka, Phys. Rev. B **81**, 081305 (2010)
27. K. Shinokita, H. Hirori, M. Nagai, N. Satoh, Y. Kadoya, K. Tanaka, Appl. Phys. Lett. **97**, 211902 (2010)
28. R. Matsunaga, N. Tsuji, H. Fujita, A. Sugioka, K. Makise, Y. Uzawa, H. Terai, Z. Wang, H. Aoki, R. Shimano, Science **345**, 1145 (2014)
29. M. Liu, H.Y. Hwang, H. Tao, A.C. Strikwerda, K. Fan, G.R. Keiser, A.J. Sternbach, K.G. West, S. Kittiwatanakul, J. Lu, S.A. Wolf, F.G. Omenetto, X. Zhang, K.A. Nelson, R.D. Averitt, Nature **487**, 345 (2012)
30. K. Itoh, H. Itoh, M. Naka, S. Saito, I. Hosako, N. Yoneyama, S. Ishihara, T. Sasaki, S. Iwai, Phys. Rev. Lett. **110**, 106401 (2013)
31. M. Mitrano, A. Cantaluppi, D. Nicoletti, S. Kaiser, A. Perucchi, S. Lupi, P. Di Pietro, D. Pontiroli, M. Ricco, S.R. Clark, D. Jaksch, A. Cavalleri, Nature **530**, 461 (2016)
32. Y. Okimoto, S. Naruse, R. Fukaya, T. Ishikawa, S. Koshihara, K. Oka, M. Azuma, K. Tanaka, H. Hirori, Phys. Rev. Applied **77**, 064016 (2017)
33. T. Miyamoto, H. Yada, H. Yamakawa, H. Okamoto, Nature com. **4**, 2586 (2013)
34. T. Umanodan, K. Kaneshima, K. Takeuchi, N. Ishii, J. Itatani, H. Hirori, Y. Sanari, K. Tanaka, Y. Kanemitsu, T. Ishikawa, S. Koshihara, S. Horiuchi, Y. Okimoto, J. Phys. Soc. Jpn. **88**, 013705 (2019)
35. S. Horiuchi, R. Kumai, Y. Tokura, J. Am. Chem. Soc. **135**, 4492 (2013)
36. S. Iwai *et.al.*, unpublished
37. T. Morimoto and S. Nagaosa, Sci. Adv. **2**, e1501524 (2016). T. Morimoto and S. Nagaosa, Phys. Rev. B **94**, 035117 (2016)

Chapter 5
Thin Film Fabrication and Novel Electronic Phases

Jun Fujioka and Yuichi Yamasaki

Abstract In this chapter, we review the recent research on the spin-state ordering in epitaxial thin films of perovskite $LaCoO_3$ from the viewpoint of optical spectroscopy and X-ray scattering. A remarkable feature is that a variety of spin/orbital orderings, which have not been identified in bulk samples, can be induced by tuning the epitaxial strain in thin film samples. In the moderately tensile-strained films, a spin-state/orbital ordering accompanying the ordering of high-spin-state site and low-spin-state site occurs around 130 K, and subsequently, a ferromagnetic/ferrimagnetic ordering emerges around 90 K. In the weakly strained films, another spin-state/orbital ordered phase with longer modulation period emerges around 40 K, followed by a ferromagnetic ordering transition around 20 K. By analyzing the Co-3*d* orbital state by the resonant soft X-ray scattering technique, we identified that two kinds of high-spin state with different site symmetries emerges in this phase. Moreover, by means of grazing-incidence resonant soft X-ray scattering technique, it is demonstrated that the transition temperature of spin-state ordering is higher in the surface-layer of film than inside the film due to the strain relaxation. These results demonstrate that the strain engineering of epitaxial thin films offers a fertile playground to study the spin-state/orbital ordering in the correlated spin-crossover material.

5.1 Introduction

The spin-state transition of the cobalt oxides has been a subject of great interest in modern material science. A remarkable feature is that both the spin quantum number (S) and orbital occupancy of Co-3*d* electrons simultaneously change upon the spin-state transition. Thanks to this unique spin-orbital entanglement, the spin-state transition manifests itself as a variety of changes in magnetic, optical, electronic,

J. Fujioka (✉)
Graduate School of Pure and Applied Sciences, University of Tsukuba, Tsukuba 305-8573, Japan
e-mail: fujioka@ims.tsukuba.ac.jp

Y. Yamasaki
National Institute for Materials Science (NIMS), Tsukuba 305-0047, Japan
e-mail: yamasaki.yuichi@nims.go.jp

Y. Okimoto et al. (eds.), *Spin-Crossover Cobaltite*, Springer Series in Materials Science 305, https://doi.org/10.1007/978-981-15-7929-5_5

and elastic properties of materials. Another important feature is that the energy scale required to induce the spin-state transition can be small in certain materials. This is essential to find novel functions originating from spin-state transitions by controlling temperature, mechanical stress, and light irradiations. One of such representative materials is the perovskite $RCoO_3$ (R = rare earth ion) [1–3]. For about four decades, $RCoO_3$ has been known as a typical correlated electron material exhibiting thermally driven spin-state transition. In this material, the nominally trivalent Co-ion (Co^{3+}) may take three different spin states; the low-spin (LS) state with filled $3d\,t_{2g}$ manifold ($S = 0$), intermediate-spin (IS) state with active e_g and t_{2g} orbital degrees of freedom ($S = 1$), and high-spin (HS) state with active t_{2g} orbital degree of freedom ($S = 2$), as shown in Fig. 5.1a. In particular, the thermally driven spin-state transition occurs at a modest temperature for the $LaCoO_3$; the crossover from the LS state to IS state and/or HS state occurs around 90 K, which manifests itself as the nonmagnetic–paramagnetic crossover [4–12]. Moreover, the thermally induced insulator–metal transition is also observed around 600 K [13–15], which has been attributed to another spin-state transition [16], whereas the specific model of this spin-state transition has remained elusive.

Another route to induce the spin-state transition is the control of lattice distortion. From the viewpoint of the single-ion model, the spin state of Co-ion is determined by the competition between the crystal field splitting between the e_g and t_{2g} manifold and the Hund's coupling. Following the ligand-field theory, the crystal field splitting is enhanced by the hybridization between the Co-3d-state and O2p-state

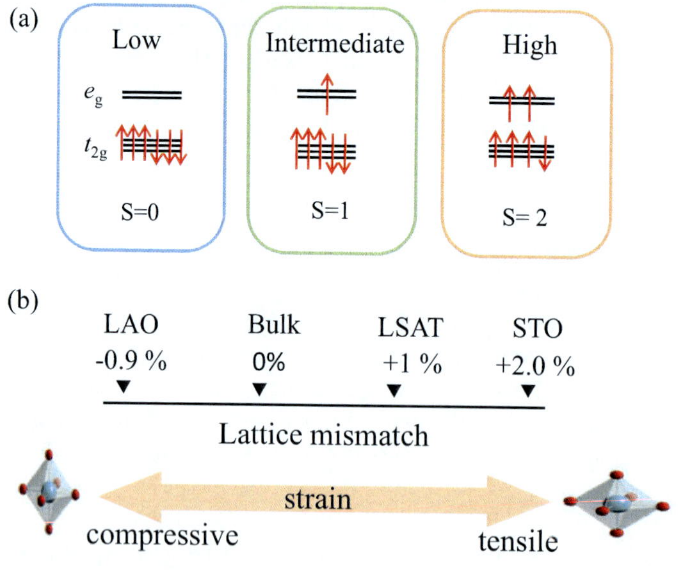

Fig. 5.1 **a** Illustration of the spin states of trivalent Co-ion (Co^{3+}): the low-spin (LS), intermediate-spin (IS), and high-spin (HS) states. The arrows denote the electron spins. **b** The relation between the lattice mismatch and epitaxial strain. The lattice mismatch is estimated by the difference between the averaged lattice constant of $LaCoO_3$ and the in-plane lattice constant of the substrate

in the isolated CoO_6-octahedra. Therefore, in this simple model, one expects that the compression (expansion) of CoO_6-octahedra results in the increase (decrease) of crystal field energy and favors the LS state (IS/HS state). A typical way to tune the lattice distortion is the application of hydrostatic pressure. Indeed, it is demonstrated that the nonmagnetic state is favored by compressing the cell volume by applying the hydrostatic pressure [17, 18]. The epitaxial strain of single-crystalline film provides is another way to control the spin state by the lattice distortion [Fig. 5.1b]. Specifically, the lattice distortion imposed by the tensile strain favors the HS state or IS state with active e_g-orbital degree of freedom, often causing the ferrimagnetism or ferromagnetism with keeping the insulating state as shown in Fig. 5.2 [19–22]. Recent studies on the basis of X-ray/electron-beam diffraction as well as the theoretical calculations suggest that the ferromagnetic/ferrimagnetic insulating state originates from the spin/orbital ordering of IS or HS state [23–26]. In general, the spin/orbital ordering is strongly coupled to the lattice distortion in perovskite oxides and thus can be significantly controlled by tuning the epitaxial strain [27–29]. Moreover, it has been also proposed that unconventional phases including the excitonic insulator can be invoked in the $LaCoO_3$ [30, 31].

One of the powerful probes to explore the spin/orbital ordering in thin films is the X-ray/electron-beam diffraction. The recent advance of synchrotron radiation science provides a variety of techniques including the resonant soft X-ray scattering and surface reflection technique in addition to the conventional hard X-ray diffraction. In particular, the resonant soft X-ray scattering is sensitive to the magnetic structure and orbital character of Co-3*d* electron, since the Co-3*d*-state is directly involved in the scattering process. Another important ingredient is the advance in the growth technique of atomically controlled single-crystalline films. Fortunately, single-crystalline films of $LaCoO_3$ can be relatively easily fabricated by using the pulse laser deposition technique. Indeed, a number of studies have already reported

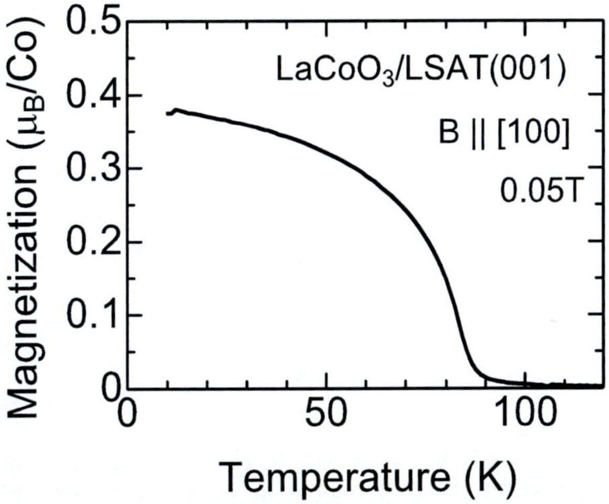

Fig. 5.2 Temperature dependence of magnetization for the $LaCoO_3$/LSAT(001) measured at 0.05 T. The data were taken from [37]

the fabrication of films with atomically smooth surface/interface or the superlattice structure combined with other perovskite oxides.

In this chapter, we review the recent advance in the spin-state transition in thin films of perovskite $LaCoO_3$. In **Sect. 5.2,** we briefly summarize the film fabrication, magnetization, and optical property. Subsequently, we present the spin/orbital ordering investigated by the hard X-ray diffraction. In Sect. 5.3, the advanced analysis of spin/orbital state probed by the resonant inelastic soft X-ray scattering is presented. Finally, we present the surface order of orbital and spin states in Sect. 5.4.

5.2 Magnetic, Structural, and Optical Property

5.2.1 Film Fabrication

The $LaCoO_3$ shows the perovskite structure with rhombohedral symmetry in a bulk form as shown in Fig. 5.3a. The single-crystalline films of $LaCoO_3$ can be fabricated on several substrates $LaAlO_3$ – (LAO), $(LaAlO_3)_{0.3}(SrAl_{0.5}Ta_{0.5}O_3)_{0.7}$ (LSAT) and $SrTiO_3$ (STO)-substrate with various crystal orientation. The illustration of the crystal structure with the (001), (110), and (111) orientations are shown in Fig. 5.3b–d. Here, the crystal orientation is denoted by the pseudo-cubic notation. The averaged lattice constants of these substrates are 3.79 Å, 3.868 Å, and 3.905 Å, respectively, in the pseudo-cubic notation. Since the averaged lattice constant of bulk $LaCoO_3$ is about

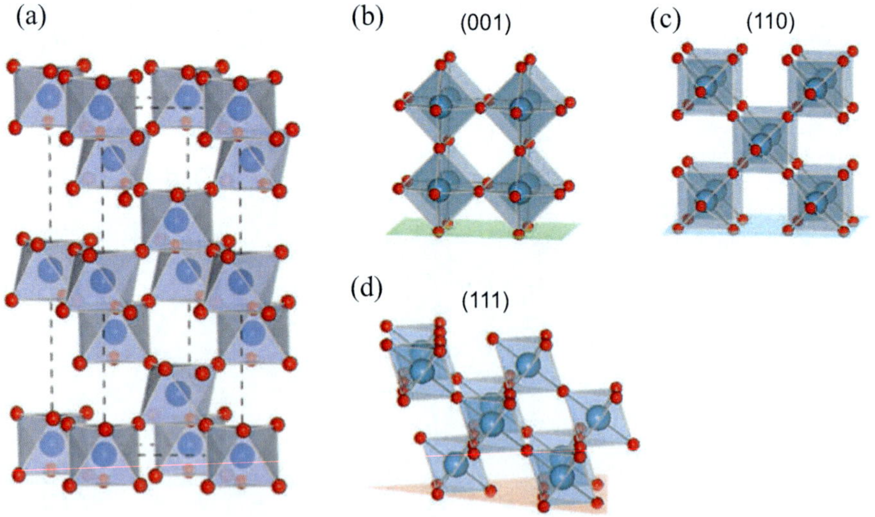

Fig. 5.3 **a** Crystal structure of bulk $LaCoO_3$ with the rhombohedral symmetry. **b–d** Schematic view of the crystal structure for the $LaCoO_3$ grown on substrates with (001), (110), and (111) orientations, respectively. The picture was taken from [37]

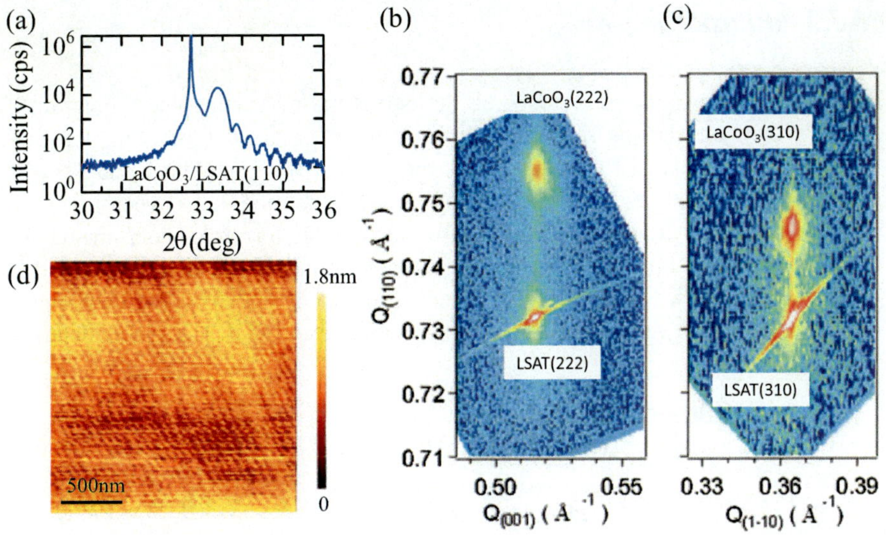

Fig. 5.4 **a** X-ray diffraction pattern at room temperature for $LaCoO_3$/LSAT(110). The sharp peak around 32.7° corresponds to the reflection of the substrate. The peak around 33.3° accompanying the Laue fringe is due to the (110) reflection of the film **b**, **c** The reciprocal space mapping around (222) and (310) reflections measured at room temperature, respectively. **d** Image of atomic force microscopy for the surface of $LaCoO_3$/LSAT(110). The stripe-like pattern corresponds to the steps and terraces of the film, suggesting the smooth surface of sample. The data were taken from Ref. [25]

3.85 Å in the pseudo-cubic notation, the compressive strain is imposed for films grown on the LAO substrate, whereas the tensile strain is imposed for films grown on LSAT and STO substrate. The strain on various substrates is shown in Fig. 5.1b.

Single-crystalline films of $LaCoO_3$ are fabricated on substrates by the pulsed laser deposition technique using a KrF excimer laser ($\lambda = 248$ nm) on various substrates. For the growth, we optimized the growth condition around 700 °C and the oxygen pressure around 0.1 mTorr. Figure 5.4a–d shows the typical X-ray diffraction patterns and image of atomic force microscopy (AFM) for films grown on LSAT (110). The X-ray diffraction pattern shows clear Laue fringes, and the secondary phase is not discernible. The AFM image shows step and terrace structures suggesting the atomically smooth surface. The averaged lattice constant of $LaCoO_3$ film, which is defined as the cubic root of the cell volume, is estimated to be 3.84 Å. This value is slightly larger (~0.3%) than that of the bulk sample, suggesting that the tensile strain is imposed. The reciprocal space mapping around (222) and (310) reflections [Fig. 5.4b, c] demonstrates that the in-plane lattice constant of the film nearly coincides with that of the substrate.

5.2.2 Phase Diagram

Figure 5.5 shows the overview of the structural/magnetic phase diagram for films grown on various substrates. In moderately strained films with the lattice mismatch larger than 1% (films grown on the LSAT(001), LSAT(110), and STO(111) substrates), the lattice distortion characterized by the modulation vector $q = (1/4, 1/4, 1/4)$ emerges around 120 K, and subsequently, the ferromagnetic/ferrimagnetic ordering emerges around 90–92 K. On the other hand, in the weakly strained thin film (~0.5%) grown on the LSAT(111) substrate, the lattice distortion characterized by the modulation vector $q = (1/6, 1/6, 1/6)$ emerges at 40 K, and subsequently, the

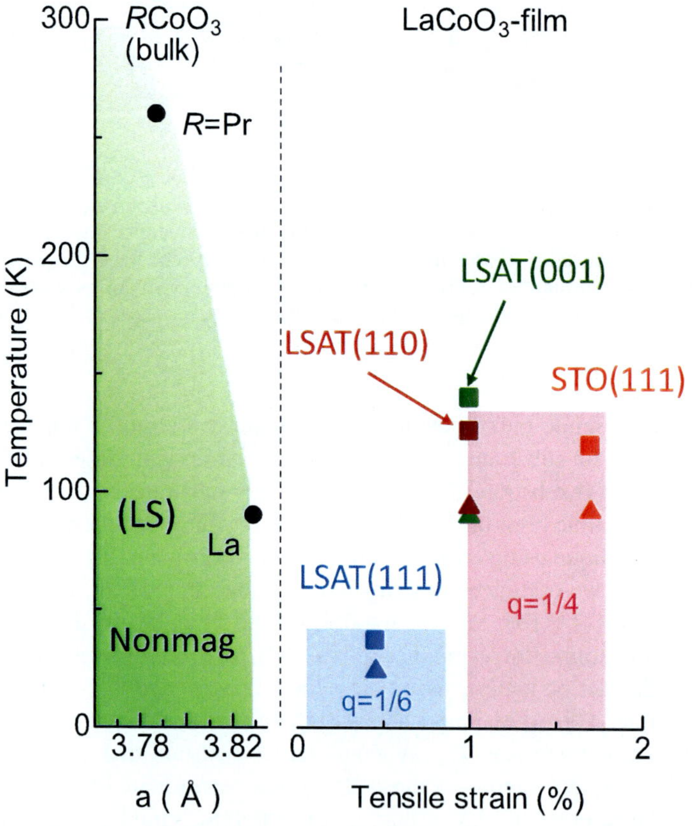

Fig. 5.5 Electronic phase diagram of the $LaCoO_3$ thin film as well as that of bulk $LaCoO_3$ and $PrCoO_3$. The vertical dashed line denotes the boundary between the phase diagram of films and that of bulk. The circle indicates the crossover temperature between nonmagnetic (NM)state and paramagnetic (PM)state. The squares and triangles denote the transition temperatures of spin-state ordering and those of ferrimagnetic (Ferri) or ferromagnetic (Ferro) ordering, respectively. q is the modulation vector of lattice distortion due to the spin-state ordering, which is denoted by the pseudo-cubic setting. The lattice constant a is defined as the cubic root of the unit cell volume in the pseudo-cubic setting. The data were taken from [37]

ferromagnetic/ferrimagnetic ordering emerges at 24 K. In the following sections, we discuss the possible origins of strain-induced ferromagnetism or ferrimagnetism on the basis of the experimental results obtained by optical spectroscopy and hard/soft X-ray diffraction.

5.2.3 Moderately Strained Films

Let us begin with a brief discussion of $LaCoO_3$/LSAT(110), which is a typical example of a moderately strained film. Figure 5.6a shows the temperature dependence of magnetization measured at 0.1 T. The onset of magnetization is observed at 94 K, which is nearly consistent with the results of $LaCoO_3$/LSAT(001). Figure 5.6b shows the magnetization curve measured at 10 K. The saturated magnetization is about 0.6 μ_B/Co. The resistivity shows an insulating behavior as shown in Fig. 5.6c and a clear charge gap of about 0.7 eV is observed by the optical conductivity spectrum [Fig.5.6d], suggesting that this material is a ferromagnetic or ferrimagnetic Mott

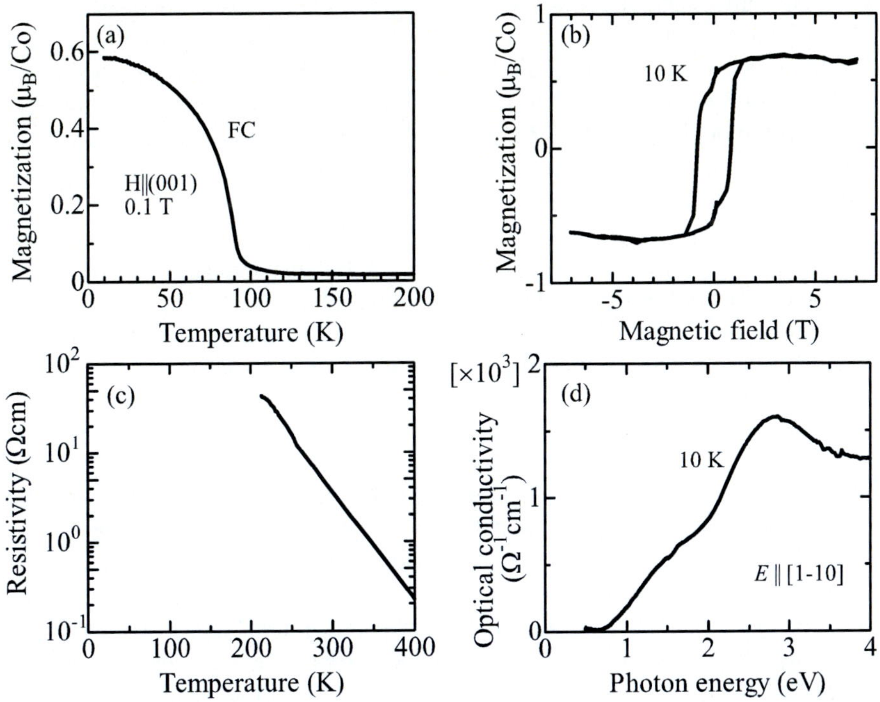

Fig. 5.6 The magnetization, resistivity, and optical conductivity for the $LaCoO_3$/LSAT(110) **a** The temperature and **b** magnetic-field dependence of magnetization. **c** Temperature dependence of resistivity. **d** The optical conductivity spectrum measured at 10 K. The data were taken from Ref. [25]

insulator. Among the possible models, the orbital ordering (or spin-state ordering) is one of the prevalent scenarios to explain the ferromagnetic or ferrimagnetic insulating phase. The orbital ordering usually accompanies the modulation of crystal structure via the collective Jahn–Teller distortion and thus can be detected by the X-ray-, neutron-, and electron-beam diffraction. On the basis of scanning transmission electron microscopy (STEM), Choi et al. report that the stripe-like lattice modulation emerges for the $LaCoO_3$/LSAT(001) and $LaCoO_3$/STO(001) [23]. The lattice modulation, which triples the unit cell of the pseudo-cubic perovskite structure, becomes remarkable as the magnitude of tensile strain is enhanced. The authors confirm the good chemical stoichiometry of films by x-ray absorption at Co L-edge and optical conductivity spectra and propose that the orbital ordering or spin state ordering of Co-3d electron is likely. Nevertheless, the oxygen vacancy is thought to be another prevalent scenario for ferromagnetism. By means of STEM, electron-energy-loss spectroscopy, and the calculation on the basis of density functional theory, Biskup et al. propose that the periodic array of oxygen vacancy induces the ordering of Co^{2+}-ion with HS state, yielding ferromagnetism/ferrimagnetism [24, 32]. The debates of these conflicting views have been unsettled at the moment, and the possible role of oxygen vacancies remains to be elusive.

On the other hand, the lattice distortion with different modulation vectors, which shows a remarkable temperature variation below room temperature, has been detected by the X-ray diffraction. In the following, we present the details of experimental results on the basis of the X-ray diffraction. Figure 5.7a shows the temperature dependence of the out-of-plane lattice constant d_{110}. The d_{110} decreases with decreasing temperature and shows a kink around 126 K. By further analyzing other Bragg reflections, it is turned out that a shear-type structural distortion along [1–10] direction occurs at 126 K ($= T_S$) as illustrated in Fig. 5.7d. Moreover, a superlattice reflection characterized by the modulation vector $q = \{1/4,1/4,1/4\}$ emerges at T_S. Figure 5.7e exemplifies the profile of superlattice reflection at (5/4 3/4 1/4). The intensity of this refection [Fig. 5.7b] gradually increases below T_S and nearly saturates below T_C. These results indicate the presence of the structural phase transition at T_S, wherein the unit cell of pseudo-cubic setting quadruples along the [100], [010], and [001] axes. The inset to Fig. 5.8a shows the scattering spectrum of (5/4 3/4 1/4) superlattice reflection as well as the absorption spectrum around Co K-edge (1 s-4p intra-atomic transition) measured at 10 K. The scattered X-ray includes both σ' and π' polarization components, while the incident X-ray is nearly polarized. Here, σ and π denote the polarization component perpendicular and parallel to the scattering plane, respectively. A clear resonant structure of the Co K-edge is seen around 7.73 keV. We have analyzed the polarization of scattered X-ray and azimuthal angle (Ψ) dependence of the scattering intensity. Here, Ψ is defined as zero, when the X-ray polarization is approximately parallel to the [−3/4, 5/4, 0] direction. Figure 5.8a shows the scattering spectra in σ → π' geometry at various Ψ. Here, the spectra are normalized by the scattering intensity of the non-resonant signal. Remarkably, even in σ → π' geometry, the resonant structure is discernible and shows an appreciable Ψ dependence [see Fig. 5.8b]. A similar periodic feature is observed in σ → σ' geometry. This indicates that the resonant scattering at the Co K-edge originates from the anisotropy

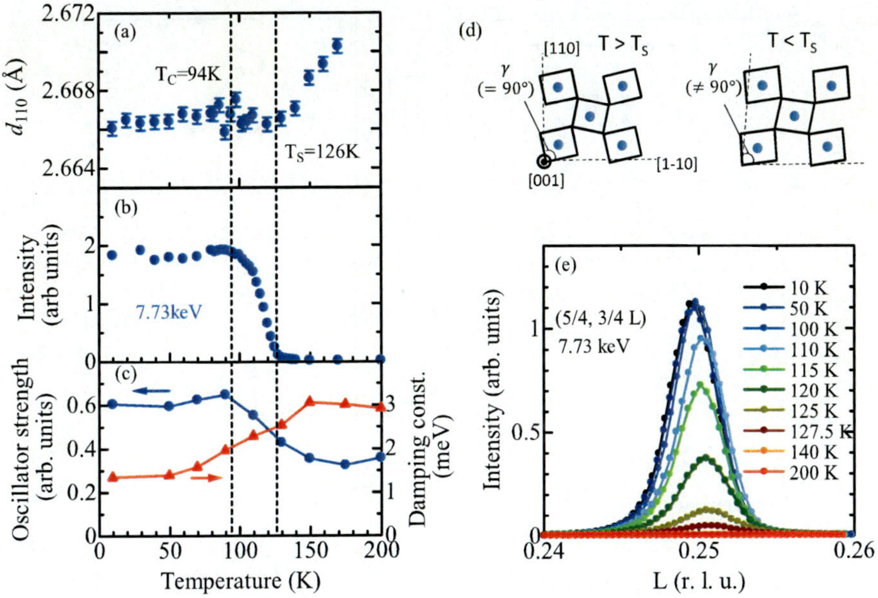

Fig. 5.7 Temperature dependence of **a** the out-of-plane lattice constant d_{110} **b** the scattering intensity of superlattice reflection measured at 7.73 keV, **c** oscillator strength (circles) and damping constant (triangles) of the Co-O stretching optical phonon. **d** Schematic view of the lattice distortion projected on the (001) plane. The dashed line is a guide to the eyes. **e** Profiles of the superlattice reflection at (5/4, 3/4, 1/4). Here, the reciprocal lattice units are represented by the basis of the LSAT substrate. r.l.u. means the reciprocal lattice unit. The data were taken from [25]

of the tensor of susceptibility [33, 34] as often observed in the correlated electron materials with orbital ordering.

Further insight into the lattice distortion can be acquired by measurements of optical phonons. Figure 5.8**c** shows energy spectra of the imaginary part of the dielectric constant ε_2, which is derived by the reflectivity spectra in the far-infrared region. At room temperature, one can see two phonon modes at 0.067 and 0.070 eV, whereas the additional two peaks grow around 0.065 and 0.072 eV below T_S. In particular, the spectral intensity of mode at 0.065 eV nearly doubles and the peak becomes sharp at low temperatures. According to [35], these infrared active phonons are assigned to the Co-O bond stretching modes, which are sensitive to the orbital state of Co-3*d* ion via the modulation of Co-O covalency. Specifically, the mode at 0.065 eV is assigned to the "Jahn–Teller mode" of CoO_6-octahedra. Such phonon activation and sharpening suggest that the Jahn–Teller distortion coupled to the Co-3*d* orbital is strongly enhanced below T_S.

Having established these results, let us consider the origin of structural phase transition. One possible scenario is the ordering of oxygen vacancies [24], which would strongly modify the environment of Co-3*d* electron via the modulation of Co-O covalency. As previously mentioned, it has been proposed that the oxygen vacancies form the superstructure characterized by a certain modulation vector in

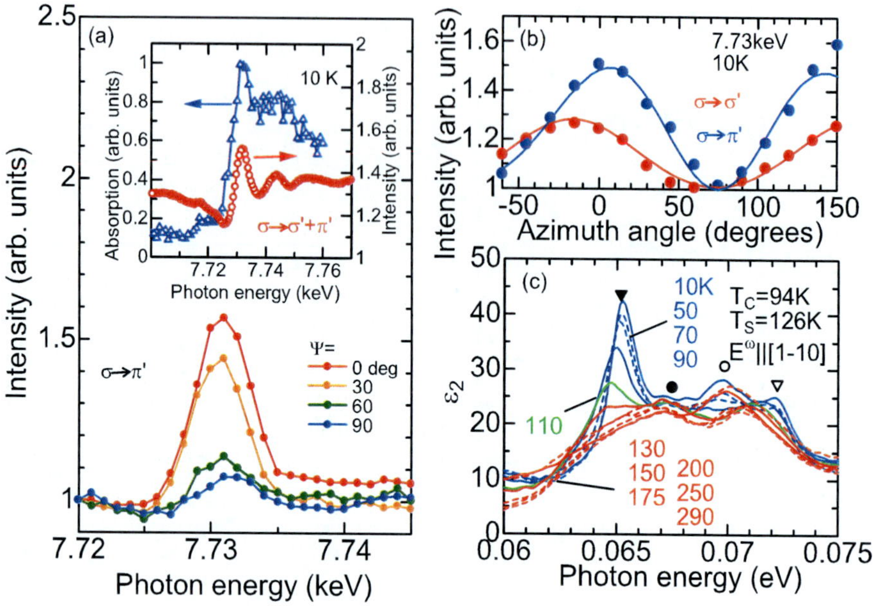

Fig. 5.8 **a** Energy spectra of the scattered intensity in the $\rightarrow \pi$ geometry. Inset is the energy spectra of the intensity of (5/4 3/4 1/4) diffraction (without polarization analysis, circles) in comparison with the x-ray absorption spectra (XAS) at Co K-edge measured in the fluorescence yield method (triangles). **b** Azimuthal angle dependence of the resonant scattering intensity measured at 10 K. The solid lines are the results of simulation of scattering intensity. **c** Energy spectra of the imaginary part of the dielectric constant (ε_2) around 0.067 eV. The peak structures marked by closed triangle, closed circle, open circle, and open triangle are assigned to the Co-O stretching mode at 0.065, 0.067, 0.070, and 0.072 eV, respectively. The data were taken from [25]

the thin film of $LaCoO_3$. However, the typical temperature required to arrange oxygen vacancies is known to be higher than 800 K in bulk $LaCoO_3$ [16, 36]. Thus, this may not be the direct origin of the structural transition at T_S (= 126 K) for the present system. Another more plausible scenario is the ordering of Co-3d-orbitals in the IS state or HS state. The orbital polarization causes the anisotropic distortion of CoO_6-octahedra and thus may manifest itself as the anisotropic tensor scattering of X-ray and anomaly of lattice dynamics. Indeed, this is consistent with the observed ψ-dependent resonant scattering at the Co K-edge and the remarkable activation of the Jahn–Teller phonons. Although the specific model of spin-orbital ordering cannot be identified from the above results alone, some plausible models could be given in Fig. 5.9a, b, including a variety of the HS-state/LS-state ordering and HS-state/IS-state ordering [10, 11, 27, 29]. The more elaborate analysis of spin-orbital ordering will be discussed on the basis of results by inelastic soft X-ray diffraction in Sect. 5.3. We note that the lattice distortion with the modulation vector of $q = \{1/4, 1/4, 1/4\}$ is also observed in the moderately strained films grown on LSAT(001) and STO(111) substrate [37]. The temperature of structural phase transition is nearly consistent with that of $LaCoO_3$/LSAT(110). Moreover, these films also show a spontaneous

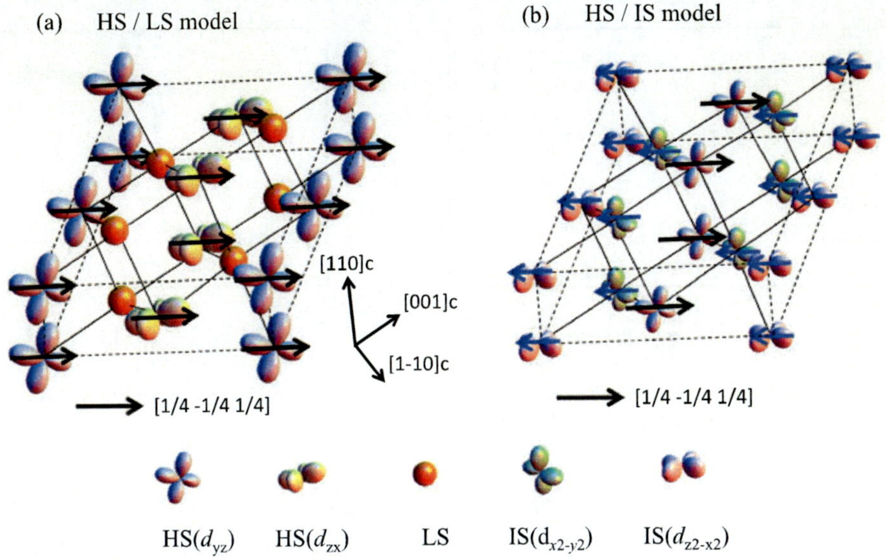

Fig. 5.9 The models of the spin and orbital ordered structure: **a** High-spin (HS)/low-spin (LS)-state-ordered state with the rocksalt-type configuration. **b** HS/Intermediate-spin (IS)-state-ordered state. The lobes denote the e_g-orbital of IS state and the active t_{2g} orbital of HS state. The fully occupied, Hund's-rule coupled t_{2g} orbital of IS state is omitted for clarity. The dashed line indicates the unit cell of this spin-orbital superstructure. The data were taken from [25]

magnetization with a similar magnetic transition temperature. The summary of the structural/magnetic phase diagram is shown in Fig. 5.5.

The spin-state ordering is also investigated by a more elaborate analysis on the lattice distortion. Sterbinsky et al. propose that the volume of CoO_6-octahedron varies with the modulation vector of (1/2, 1/2, 1/2) by means of the X-ray reflection experiment and the theoretical calculation on the basis of the density functional theory [26]. They analyzed the resonant x-ray reflection of the *K*-edge of Co-ion and concluded that the octahedral breathing distortion coupled to the HS-state/LS-state ordering occurs in the film grown on STO(001) substrate.

5.2.4 Weakly Strained Films

Figure 5.10a shows the temperature dependence of magnetization for $LaCoO_3$ thin films grown on LSAT(111) substrates [37]. In this material, the ferromagnetic or ferrimagnetic transition occurs at 24 K (= T_C). The magnetization curve shows the saturated magnetization of about 0.7 μ_B/f.u and the coercive field is about 0.03 T [Fig. 5.10b]. The T_C and coercive field are much smaller than those of $LaCoO_3$/LSAT(110), suggesting the emergence of different kinds of spin-orbital ordering. In particular, the out-of-plane lattice constant d_{111} shows a non-monotonic

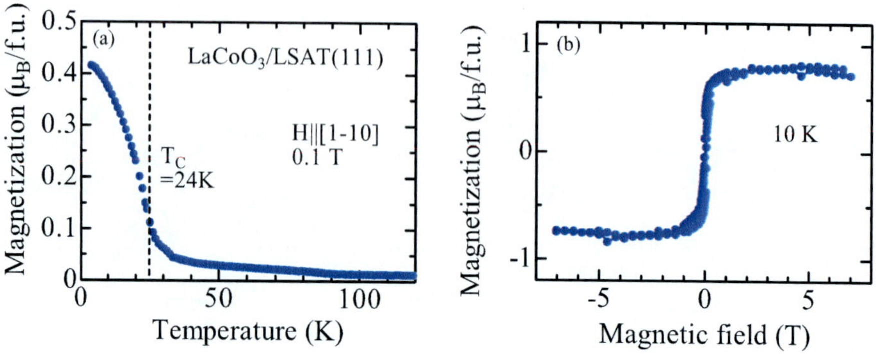

Fig. 5.10 **a** The temperature dependence of magnetization and **b** the magnetization curve measured at 10 K for $LaCoO_3$/LSAT(111). The data were taken from [37]

temperature dependence and becomes minimum at around 50 K as shown in Fig. 5.11a. The spectral shape of the optical phonon for the ε_2-spectra is also different from that of $LaCoO_3$/LSAT(110) and significantly changes below 30 K; a new mode appears around 69 meV [Fig. 5.11b, d]. In addition, the superlattice reflection with the modulation vector $q = (1/6, 1/6, 1/6)$ appears below 40 K as shown in Fig. 5.11e. These results suggest that a structural phase transition, in which the lattice constant of the pseudo-cubic perovskite structure becomes 6 times in the [100], [010], and [001] directions, occurs at around 40 K. Figure 5.12 shows the scattering spectra of the Co K-edge at various azimuthal angles Ψ around [111] axis. Here, the incident beam is nearly σ-polarized, but the scattered beam contains both the σ- and π-components. Unlike the case of $LaCoO_3$/LSAT(110), the spectra do not show a significant Ψ dependence, suggesting that the local lattice symmetry around the Co site is almost isotropic with respect to rotation around the [111] direction. This is one of the important ingredients to consider the possible model of orbital ordering of the present system. In the model, which includes the ordering of e_g-orbital such as dx^2-y^2 and dz^2, the Jahn–Teller distortion would cause sizeable anisotropy around [111] axis, unless the order parameter is extremely small. In this regard, it is more likely to consider the model, which constitutes the ordering of LS and/or HS without the active e_g-orbital degree of freedom.

Figure 5.13 exemplifies the possible models of spin-orbital ordering. For clarity, we also show the schematic view of the crystal structure of bulk $LaCoO_3$ with the rhombohedral symmetry in Fig. 5.13a. Figure 5.13b shows the HS/LS disproportionation model, wherein the HS-state sites align in every six sites along the [111] direction in the matrix of LS state. Here, we assumed that the lowest-energy state is $J_{\text{eff}} = 1$ triplet, taking into account the moderate spin-orbit coupling under the weak trigonal crystal field (the compressive lattice distortion along [111] direction). The expected net magnetization of this model is about $0.57\mu_B$/Co, given that the spin of HS-site with local moment ($3.4\mu_B$/Co) [5, 6] is parallel to each other. This value nearly agrees with the observed saturated magnetization. Another possible model is the HS/LS disproportionation model composed of LS-state sites and two

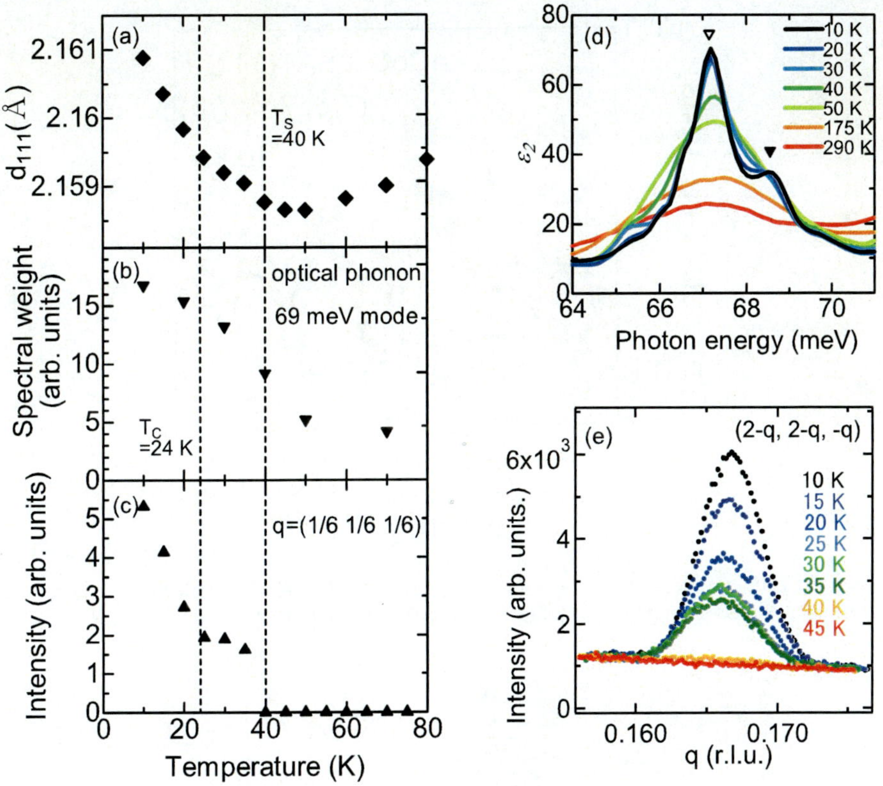

Fig. 5.11 Temperature dependence of **a** out-of-plane lattice constant d_{111}, **b** spectral intensity of the Co-O stretching optical phonon at 69 meV, and **c** scattering intensity of superlattice reflection at (11/6, 11/6, −1/6). **d** Energy spectra of the imaginary part of the dielectric constant (ε_2) for the Co-O bond stretching phonon. The peak structure marked by open (closed) triangles denotes the mode at 67 meV (69 meV). **e** Profiles of superlattice reflection at (11/6, 11/6, −1/6) measured at 12 keV. The data were taken from [37]

inequivalent HS-state sites as illustrated in Fig. 5.13c. We show two examples; the HS-state sites with inequivalent valence state ($Co^{3-\delta}$ and $Co^{-(3-\delta)}$) or those with inequivalent hybridization with LS-state $\alpha_1|HS + \beta_1|LS$ and $\alpha_2|HS + \beta_2|LS$ with $= \alpha_1 \neq \alpha_2, \beta_1 \neq \beta_2$ [30, 31]. We note that the above considerations are limited to the cases where the spin configurations are collinear, and there may be other more reasonable models with noncollinear spin ordering. Further analysis on the basis of resonant soft X-ray scattering will be described in the following sections.

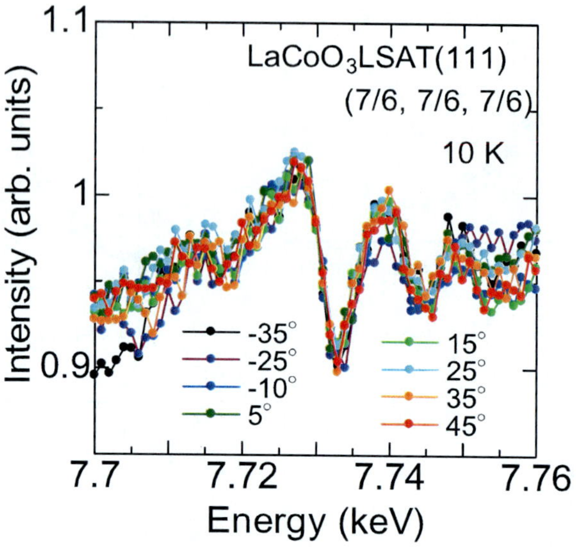

Fig. 5.12 Energy dependence of resonant intensity at (7/6, 7/6, 7/6) reflection. ψ denotes the azimuthal angles around the [111] direction. The ψ is defined as zero, when the incident

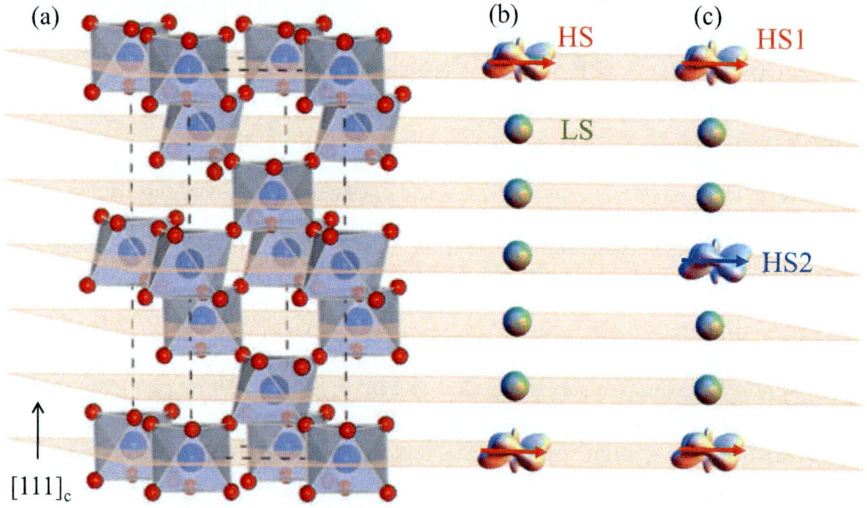

Fig. 5.13 **a** Crystal structure of $LaCoO_3$ with the rhombohedral symmetry. Dashed lines indicate the unit cell. Note that the [001] axis in the rhombohedral setting corresponds to the [111] axis in the pseudo-cubic setting. **b** Schematic view of spin-state ordering including the low-spin (LS) state sites and high-spin (HS) state sites. The lobes exemplify the $J_{\mathrm{eff}} = 1$, $J_z = \pm 1$ state of the HS state. Spheres represent the LS-state sites with fully occupied t_{2g} manifold. The ordered spins are denoted with arrows. **c** Schematic view of spin-state ordering including the LS-state site and two inequivalent HS-state sites. The data were taken from [37]

5.3 Spin State of $LaCoO_3$ Observed via Resonant Inelastic Soft X-Ray Scattering

In previous sections, the spin states of Co^{3+} ion in $LaCoO_3$ thin films were estimated from the spatial orderings of 3*d* electron states via resonant x-ray diffraction techniques. However, direct observations of the electronic structures are required to clarify the spin states of Co^{3+} ion. The HS/LS ratio is hard to determine by conventional methods such as x-ray absorption spectroscopy (XAS), whereas resonant *inelastic* soft x-ray scattering (RIXS) by using Co $2p \rightarrow 3d \rightarrow 2p$ process (L-edge) is a powerful technique to investigate the spin states via observing the *d-d* excitations [38–40].

Figure 5.14**a** shows XAS spectra of $LaCoO_3$ thin films fabricated on LSAT(110) and LSAT(111) substrates at Co *L*-edge measured by the total electron yield method. The peaks around 779 eV and 794 eV correspond to the Co L_3 and L_2 edges, respectively. Although the magnitude of the tensile strain and modulation vector is different between $LaCoO_3$ thin films on LSAT(110) and LSAT(111), the spectral shapes are quite similar, which indicates that it is difficult to distinguish the spin state only from

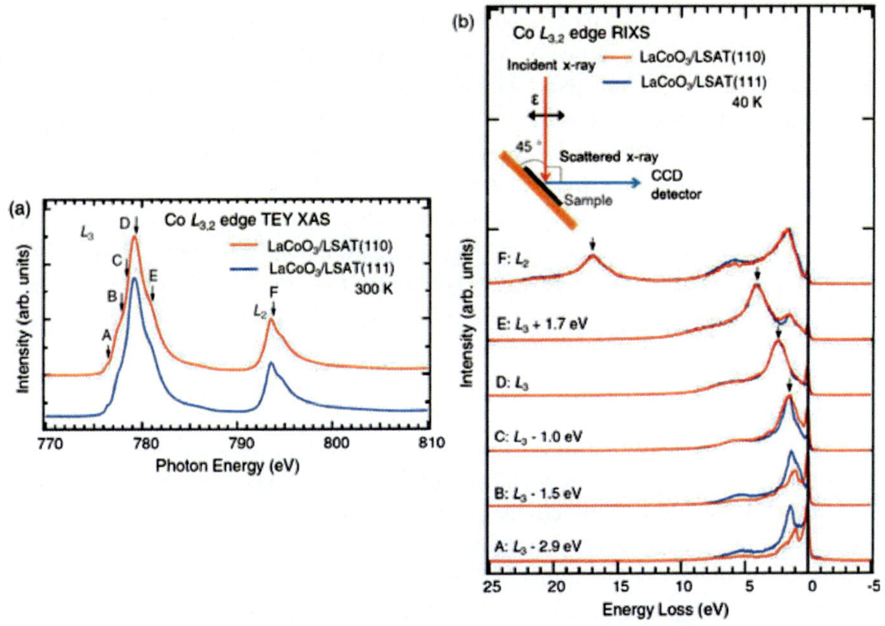

Fig. 5.14 Spectra measured at the Co L_3, L_2 edges by **a** the total electron yield x-ray absorption (TEY XAS) and **b** resonant inelastic x-ray scattering (RIXS) spectra of $LaCoO_3$ thin films grown on LSAT (110) and LSAT(111) substrates. The arrows and letters in XAS indicate the excitation energies for RIXS. All RIXS spectra are normalized by the intensity of the highest peak and the arrows indicate the fluorescence peaks. The inset shows the schematic diagram of the experimental setup in our RIXS measurements. The data were taken from [41]

XAS. Whereas differences in spectra between those two thin films are clearly seen in RIXS spectra as shown in Fig. 5.14**b** [41]. Labels A–F indicate the energies of incident soft x-ray, and the inset shows the layout of the RIXS experiment performed at BL07LSU HORNET, SPring-8 [42]. In the range of the x-ray energy, the energy resolution is about 300 meV. The charge coupled device (CCD) detector was set at 90 degrees relative to the incident x-ray with horizontal polarization to suppress the elastic scattering.

Peaks from 0 eV to 4 eV should correspond to the *d-d* excitations and the peak around 5.0 eV may be a charge-transfer (CT) excitation. Whereas the peaks seen at around 0 eV are coming from the elastic scattering and peaks indicated by arrows correspond to the fluorescence yield. In the spectra of A and B, the difference in *d-d* excitation peaks is clearly observed between $LaCoO_3$/LSAT(110) and $LaCoO_3$/LSAT(111), indicating that the spin states of $LaCoO_3$ change drastically according to the magnitude of the tensile strain. Whereas the peaks of *d-d* excitation peaks are not clear in the spectra excited by higher energy (C, D, E, and F), which might be ascribed to the influence from larger fluorescence yield peaks.

To estimate the spin state from the RIXS spectra, it is necessary to compare the experimental results with the RIXS spectra obtained by theoretical calculations. Figure 15**a** shows the RIXS spectra with the excitation energy A: L_3-2.9 eV for two thin films and a bulk crystal of $LaCoO_3$ measured at 40 K and 300 K, comparing with the spectra obtained by impurity Anderson model calculations [41]. There are two kinds of strong peaks in the energy range between 0 and 2 eV in all three samples. As increasing temperature, the behavior that the peak intensities at 0.3 eV increase and those at 1.3 eV decrease for RIXS spectra of bulk crystal should correspond to the fact that the population of the HS state increases with increasing temperature. This indicates that the peaks observed around 0.3 eV can be assigned to the excitations from the HS ground states, whereas the peaks around 1.3 eV correspond to the LS ground states. The spectra for $LaCoO_3$/LSAT(111) are similar to those for bulk; however, peaks that appear at 1.0 eV in $LaCoO_3$/LSAT(110) can be explained by neither the LS nor HS states.

The calculations are performed for the HS state and the LS state with O_h site symmetry, and the HS state with distorted D_{2h} site symmetry. As seen in the theoretical spectra, the peak for HS state is shifted to 1.0 eV by lowering the symmetry from O_h to D_{2h}, indicating that the spin state of $LaCoO_3$/LSAT(110) consists of the HS states with lower local symmetries, *i.e.,* the mixture of O_h and D_{2h} symmetries. By comparing those experimental and theoretical spectra, the ratio of spin states can be estimated by linear combinations of the theoretical spectra. As shown in Fig. 5.15**b**, the combination of theoretical spectra at HS(O_h): HS(D_{2h}) = 1: 1 agrees well with the experimental spectrum of $LaCoO_3$/LSAT(110). This suggests that the CoO_6-octahedra in O_h and D_{2h} local symmetry coexist as the ratio of 1:1. Considering the fourfold periodic spatial modulation of electronic state in $LaCoO_3$/LSAT(110), one plausible model is that the spin states are aligned in order of O_h - D_{2h} (d_{yz}) - O_h - D_{2h} (d_{zx}), where the HS states with D_{2h} have two different orbital states. We note that this model is equivalent to the one shown in Fig. 5.9**a** with the LS-site replaced by the HS(O_h)-site. On the other hand, the combination of theoretical spectra at

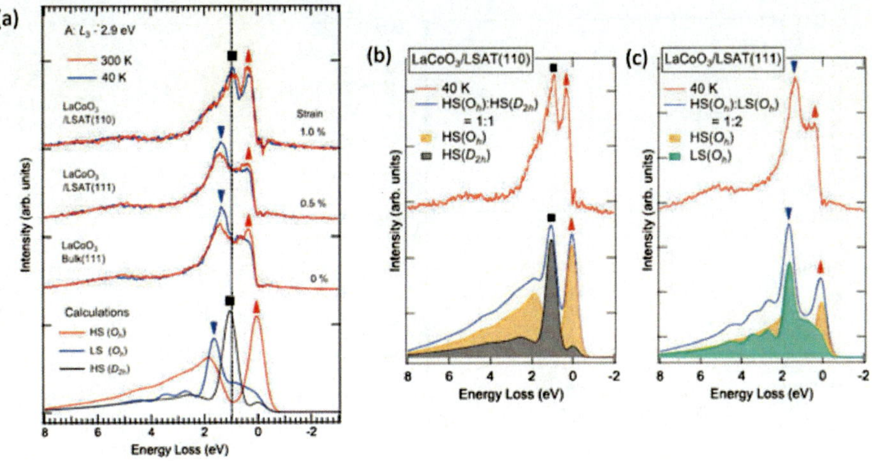

Fig. 5.15 **a** Temperature dependence (300 and 40 K) of the RIXS spectra excited at A: L3 − 2.9 eV (776.5 eV) in comparison with the impurity Anderson model calculations in O_h and D_{2h} symmetries. The triangles, squares, and inverted triangles indicate the main peak of the HS(O_h), HS(D_{2h}), and LS(O_h) states, respectively. Comparison between the experimental RIXS and the linear combinations of theoretical spectra in (**b**) $LaCoO_3$/LSAT(110) and **b** $LaCoO_3$/LSAT(111). The data were taken from [41]

HS(O_h): LS(O_h) = 1: 2 shows good agreement with the experimental spectrum of $LaCoO_3$/LSAT(111) as shown in Fig. 5.15**c**. The result is consistent with the threefold periodic modulation as observed via resonant x-ray scattering in $LaCoO_3$/LSAT(111) as illustrated in Fig. 5.13**c**.

5.4 Surface Order of Orbital and Spin State in $LaCoO_3$ Thin Film

Up to the previous section, we have seen that the orbital and spin-state order appear in the tensile-strained thin films, but it has been observed that different electronic states are realized in $LaCoO_3$/LSAT(110) thin film sample inside and on the surface [43]. As discussed in Sect. 5.2.3, the thin film shows the ferrimagnetic order below $T_S = 95$ K, while the fourfold periodic spin state and orbital order shows up below $T_O = 115$ K. Since the spatial modulation vector is parallel to the surface, a grazing-incidence (GI) condition where the incident angle is tuned to be near or below the critical angle of total reflection is required to observe the diffraction via resonant soft x-ray scattering (RSXS). At the grazing angle, the incident soft x-ray evanescent wave is confined to a depth of a few nm beneath the surface. Hence, resonant x-ray scattering at hard x-ray of Co *K*-edge ($\sim$ 7.7 keV) can detect the modulation structure for the entire thin film, while that at soft x-ray of Co *L*-edge observes electronic structure only on the sample surface.

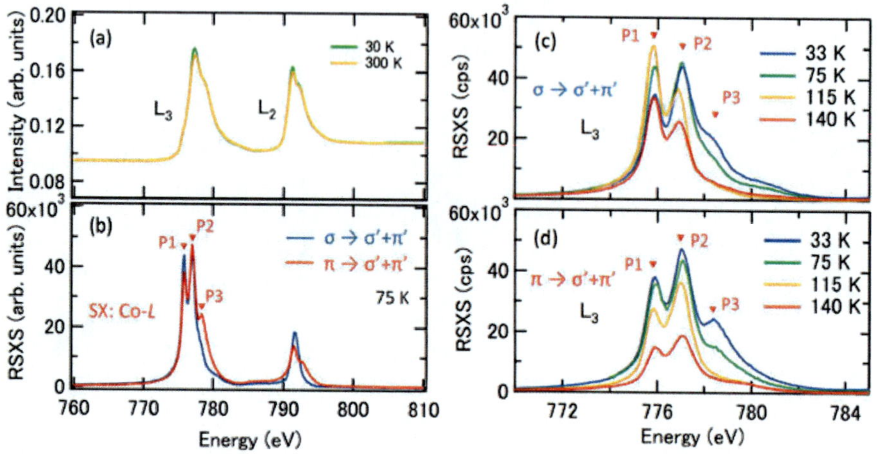

Fig. 5.16 **a** Soft x-ray absorption spectrum (XAS) measured using the partial fluorescence yield in the reflection condition. **b** Grazing-incident resonant soft x-ray scattering (GI-RSXS) spectra for $\sigma \rightarrow \sigma' + \pi'$ and $\pi \rightarrow \sigma' + \pi'$ processes at 75 K. Temperature variation of GI-RSXS spectra for **c** $\sigma \rightarrow \sigma' + \pi'$ and **d** $\pi \rightarrow \sigma' + \pi'$ processes. The data were taken from [43]

Figure 5.16a, b shows the x-ray absorption spectra and the GI-RSXS spectra, respectively. At the L_3 edge, there are three resonant scattering peaks observed at 776 eV (P1), 777 eV (P2), and 778 eV (P3). In Fig. 5.16c, d, the RSXS profiles at L_3 edge are displayed for σ and π polarization of incident soft x-rays, respectively, as a function of temperature. Peak intensity for P3 suppresses above the magnetic ordering temperature of $T_S = 95$ K, whereas those for P1 and P2 survives up to 140 K. The behavior indicates that the P3 is mainly of magnetic origin, while the P1 and P2 can be ascribed to the orbital and/or spin-state order. However, those peaks appear to survive even above the orbital ordering temperature of $T_O = 115$ K determined by the bulk-sensitive resonant x-ray scattering.

To check the details of the transition temperature of magnetic and orbital ordering, Fig. 5.17**a**, **b** shows the temperature dependence of P1 and P3 with σ polarization in comparison with that of magnetization and intensity of orbital order observed by resonant x-ray scattering at Co K-edge. As increasing temperature, the intensity of magnetic diffraction P3 is suppressed above 75 K, which is about 20 K lower than the onset of magnetization ($T_S = 95$ K). The P1 and P3 finally disappear at 145 K, which is about 30 K higher than $T_O = 115$ K. Those results suggest the coexistence of orbital ordered phases with different transition temperatures. Taking into account the probing depth, one can conclude that GI-RSXS comes from the electronically ordered states emerging at the surface, where the transition temperature becomes higher than the bulk. In other words, the orbital is ordered on the surface but disordered in the bulk in the temperature range between 115 K and 145 K.

Those two orbital ordered phases inside and on the surface of the thin film can be simultaneously observed at the superlattice reflection at $(2 - q, 1, 1 - q)$ with $q \sim 1/4$ by hard x-ray diffraction. As shown in Fig. 5.18a, the superlattice peak

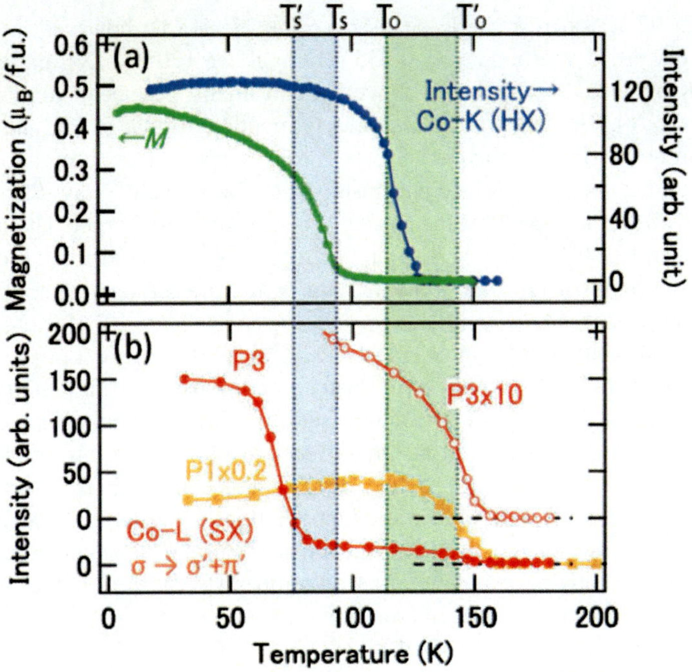

Fig. 5.17 Temperature dependence of **a** magnetization (M) and intensity of superlattice peaks at (2 + q, 2 − q, q) with q = 1/4 measured by hard x-ray diffraction (HX) and **b** intensities of superlattice peaks at (q, − q, q) for P1 and P3 of GI-RSXS (SX). The data were taken from [43]

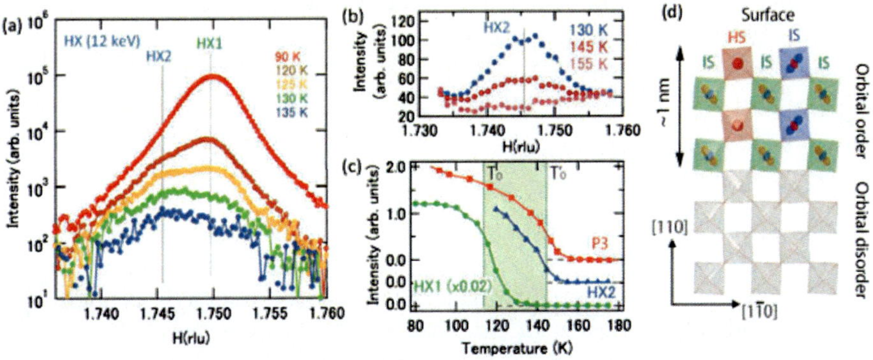

Fig. 5.18 **a**, **b** Temperature dependence of peak profile at (2 − q, q, 1 − q) reflection measured by non-resonant hard x-ray (12 keV) diffraction, in **a** logarithmic and **b** linear scales. To probe the weak intensity of the superlattice reflection at higher temperatures, the integral times for the data in (b) are one hundred times longer than those in **a**. **c** Temperature dependence of the intensities for the reflections of HX1 (of bulk origin) and HX2 (of surface origin) in comparison with the GI-RSXS result [Fig. 5.17**b**]. **d** Schematic image of the surface order and bulk disorder of the orbital degree of freedom, as realized in the temperature region between T_O and T'_O. The data were taken from [43]

profile at 90 K forms a single peak at $q = 0.25$ (HX1); however, an additional peak is discerned at around $q = 0.255$ (HX2) above 120 K. HX1 almost vanishes at 125 K, whereas HX2 subsists above T_O and finally disappears almost at 155 K [Fig. 5.18b, c]. The onset of HX2 is nearly identical to that of P3, indicating that HX2 can be ascribed to the orbital order at the surface. The observation of phase separation between the surface and inside thin film by hard x-ray diffraction thus provides further evidence for the surface freezing of orbital order. The intensity of HX2 is merely 1.6% of that of HX1, indicating that the surface orbital order appears only at the topmost layer of about 1.0 nm, which corresponds to four Co layers [Fig. 5.18d].

There is no lattice relaxation in $LaCoO_3$/LSAT(110), thus the observed surface freezing of orbital order should be intrinsic in origin, perhaps relevant to the broken translational/inversion symmetry at the surface. Structural instability sometimes causes a lattice reconstruction at the surface. In the present case of $LaCoO_3$, if the distortion of CoO_6 cluster, i.e., D_{2h} distortion, is enhanced at the surface, it will lead to stabilize the four periodic orbital modulations at the surface. The reconstruction of orbital order should also influence the magnetic exchange interactions, which may result in changing magnetic ordering temperature at the surface. As a result, the novel phenomenon of surface ordering and bulk disordering of the orbital degree of freedom is realized in $LaCoO_3$/LSAT(110) thin film, which can be attributed to its collective and lattice-coupled nature which is strongly influenced by the translational/inversion symmetry breaking at the surface. Conversely, the phenomenon of melting orbital order only on the surface has been observed in layered manganese oxides [44, 45]. The surface ordering temperature should be sensitively determined depending on the ordering structure of the electronic state and the relaxation of the lattice structure on the surface.

5.5 Concluding Remarks

In this chapter, we reviewed recent experimental research in terms of magnetization measurement, infrared spectroscopy, and X-ray diffraction. We showed that the ferromagnetic ordering or ferrimagnetic ordering is strongly coupled to the spin-state/orbital ordering of Co-3*d* electron in the tensile-strained thin films of $LaCoO_3$. From the analysis of lattice distortion by hard X-ray diffraction, we found that two kinds of the spin-state/orbital ordered phases emerge in these films; one with the lattice modulation vector of {1/4, 1/4, 1/4} in the moderately strained case [$LaCoO_3$/LSAT(110)] and another with the lattice modulation vector of {1/6, 1/6, 1/6} in the weakly strained case [$LaCoO_3$/LSAT(111)]. From the inelastic soft X-ray diffraction results and theoretical calculation, it is likely that the ordered phase includes the two kinds of HS-site with O_h-local lattice symmetry and D_{2h}-one in the moderately strained case. Considering the lattice modulation of {1/4, 1/4, 1/4}, we anticipate that the spin states are aligned in order of O_h - D_{2h} (d_{yz}) - O_h - D_{2h} (d_{zx}), where the HS states with D_{2h} have two different orbital states. On the other hand, in the weakly strained case, the model of HS(O_h): LS(O_h) = 1: 2 shows good agreement with the experimental spectrum in the weakly strained case. The ordered phase may be characterized by the threefold periodic array of HS(O_h) and LS(O_h), although further investigation is required to refine the spin state in this case.

Besides the "bulk" of films, we also described that the surface shows a distinct ordered state. By using the grazing-incident resonant X-ray scattering, we found that the orbital ordering subsists up to 145 K at the surface of $LaCoO_3$/LSAT(110). This temperature is about 20–30 K higher than the transition temperature of "bulk" film. From the analysis of superlattice reflection, the surface ordering occurs about the topmost layer of about 1.0 nm, which amounts to a few atomic layers. We anticipate that the broken translational/inversion symmetry at the surface plays an important role to realize such a surface state, although there may be some contribution from the chemical modulation inherent to the surface such as the oxygen vacancy. In any case, this suggests that the orbital state of this material is sensitive to the lattice symmetry or chemical modulation and thus may offer a fertile playground to develop a novel orbital state and magnetism at the surface or interface.

At the time of writing this section, the investigations of spin/orbital ordering in the thin film of $LaCoO_3$ are still in progress. There are several rational models of spin-orbital ordering consistent with the experimental results. Further analysis/experiments are required to settle the discussion. In addition, there are few works about the low energy dynamic of the spin/orbital ordering. The future experimental and theoretical investigation may clarify the novel aspect of this old and new material.

Acknowledgements We thank H. Nakao, R. Kumai, J. Okamoto, T. Sudayama, R. Fukaya, Y. Murakami, M. Nakamura, A.Doi, M. Kawasaki, T. Arima, Y. Tokura, M. Taguchi, Y. Hirata, K. Takubo, J. Miyawaki, Y. Harada, D. Asakura, H. Daimon, Y. Yokoyama, and H Wadati for collaboration on work related to this review and for helpful discussions.

References

1. P.M. Raccah, J.B. Goodenough, First-Order Localized-Electron Collective-Electron Transition in $LaCoO_3$. Phys. Rev. B **155**, 932 (1967)
2. V.G. Bhide, D.S. Rajoria, G.R. Rao, C.N.R. Rao, Mossbauer Studies of the High-Spin-Low-Spin Equilibria and the Localized-Collective Electron Transition in $LaCoO_3$. Phys. Rev. B **6**, 1021 (1972)
3. R.R. Heikes, R.C. Miller, R. Mazelsky, Magnetic and electrical anomalies in $LaCoO_3$. Physica **30**, 1600 (1964)
4. M.A. Korotin et al., Intermediate-spin state and properties of $LaCoO_3$. Phys. Rev. B **54**, 5309 (1996)
5. C. Zobel et al., Evidence for a low-spin to intermediate-spin state transition in $LaCoO_3$. Phys. Rev. B **66**, 020402(R) (2002)
6. S. Noguchi, S. Kawamata, K. Okuda, H. Nojiri, M. Motokawa, Evidence for the excited triplet of Co^{3+} in $LaCoO_3$. Phys. Rev. B **66**, 094404 (2002)

7. K. Knizek, Z. Jirak, J. Hejtma´nek, M. Veverka, M. Marysko, G. Maris, and T. T. M. Palstra, Structural anomalies associated with the electronic and spin transitions in $LnCoO_3$. Eur. Phys. J. B 47, 213 (2005)
8. A. Podlesnyak, S. Streule, J. Mesot, M. Medarde, E. Pomjakushina, K. Conder, A. Tanaka, M.W. Haverkort, D.I. Khomskii, Spin-State Transition in $LaCoO_3$: Direct Neutron Spectroscopic Evidence of Excited Magnetic States. Phys. Rev. Lett. **97**, 247208 (2006)
9. M.W. Haverkort, Z. Hu, J.C. Cezar, T. Burnus, H. Hartmann, M. Reuther, C. Zobel, T. Lorenz, A. Tanaka, N.B. Brookes, H.H. Hsieh, H.-J. Lin, C.T. Chen, L.H. Tjeng, Spin State Transition in $LaCoO_3$ Studied Using Soft X-ray Absorption Spectroscopy and Magnetic Circular Dichroism. Phys. Rev. Lett. **97**, 176405 (2006)
10. K. Knizek, Z. Jirak, J. Hejtmanek, P. Novak, W. Ku, GGA + U calculations of correlated spin excitations in $LaCoO_3$. Phys. Rev. B. **79**, 014430 (2009)
11. J. Kunes, V. Krapek, Disproportionation and metallization at low-spin to High-Spin Transition in Multiorbital Mott Systems. Phys. Rev. Lett. **106**, 256401 (2011)
12. A. Doi, J. Fujioka, T. Fukuda, S. Tsutsui, D. Okuyama, Y. Taguchi, T. Arima, A.Q.R. Baron, Y. Tokura, Multi-spin-state dynamics during insulator-metal crossover in $LaCoO_3$. Phys. Rev. B **90**, 081109(R) (2014)
13. K. Asai, A. Yoneda, O. Yokokura, J.M. Tranquada, G. Shirane, K. Kohn, Two spin-state transitions in $LaCoO_3$. J. Phys. Soc. Jpn. **67**, 290 (1998)
14. Y. Tokura, Y. Okimoto, S. Yamaguchi, H. Taniguchi, T. Kimura, H. Takagi, Thermally induced insulator-metal transition in $LaCoO_3$: a view based on the Mott transition. Phys. Rev. B **58**, R1699 (1998)
15. S. Yamaguchi, Y. Okimoto, Y. Tokura, Bandwidth dependence of insulator-metal transitions in perovskite cobalt oxides. Phys. Rev. B **54**, R11022 (1996)
16. P.G. Radaelli, S.W. Cheong, Structural phenomena associated with the spin-state transition in $LaCoO_3$. Phys. Rev. B **66**, 094408 (2002)
17. K. Asai, O. Yokokura, M. Suzuki, T. Naka, T. Matsumoto, H. Takahashi, N. Mori, K. Kohn, Pressure Dependence of the 100 K Spin-State Transition in $LaCoO_3$. J. Phys. Soc. Jpn. **66**, 967 (1997)
18. T. Vogt, J.A. Hriljac, N.C. Hyatt, P. Woodward, Pressure-induced intermediate-to-low spin state transition in $LaCoO_3$. Phys. Rev. B **67**, 140401(R) (2003)
19. D. Fuchs, C. Pinta, T. Schwarz, P. Schweiss, P. Nagel, S. Schuppler, R. Schneider, M. Merz, G. Roth, and H. v. Lohneysen, Ferromagnetic order in epitaxially strained $LaCoO_3$ thin films. Phys. Rev. B 75, 144402 (2007)
20. J.W. Freeland, J.X. Ma, J. Shi, "Ferromagnetic spin-correlations in strained $LaCoO_3$ thin films" Appl. Phys. Lett. **93**, 212501 (2008)
21. V.V. Mehta, M. Liberati, F.J. Wong, R.V. Chopdekar, E. Arenholz, Y. Suzuki, Ferromagnetism in tetragonally distorted $LaCoO_3$ thin films. J. Appl. Phys. **105**, 07E503 (2009)
22. A.D. Rata, A. Herklotz, L. Schultz, K. Dorr, Lattice structure and magnetization of $LaCoO_3$ thin films. Eur. Phys. J. B **76**, 215 (2010)
23. W.S. Choi, J.-H. Kwon, H. Jeen, J.E. Hamann-Borrero, A. Radi, S. Macke, R. Sutarto, F. He, G.A. Sawatzky, V. Hinkov, M. Kim, H.N. Lee, Strain-Induced Spin States in Atomically Ordered Cobaltites. Nano Lett. **12**, 4966 (2012)
24. N. Biskup, J. Salafranca, V. Mehta, M.P. Oxley, Y. Suzuki, S.J. Pennycook, S.T. Pantelides, M. Varela, Phys. Rev. Lett. **112**, 087202 (2014)
25. J. Fujioka, Y. Yamasaki, H. Nakao, R. Kumai, Y. Murakami, M. Nakamura, M. Kawasaki, Y. Tokura, Spin-Orbital Superstructure in Strained Ferrimagnetic Perovskite Cobalt Oxide. Phys. Rev. Lett. **111**, 027206 (2013)
26. G.E. Sterbinsky, R. Nanguneri, J.X. Ma, J. Shi, E. Karapetrova, J.C. Woicik, H. Park, J.-W. Kim, P.J. Ryan, Ferromagnetism and Charge Order from a Frozen Electron Configuration in Strained Epitaxial $LaCoO_3$. Phys. Rev. Lett. **120**, 197201 (2018)
27. H. Hsu, P. Blaha, R.M. Wentzcovitch, Ferromagnetic insulating state in tensile-strained $LaCoO_3$ thin films from LDA + U calculations. Phys. Rev. B **85**, 140404(R) (2012)

28. J.M. Rondinelli, N.A. Spaldin, Structural effects on the spin-state transition in epitaxially strained $LaCoO_3$ films. Phys. Rev. B **79**, 054409 (2009)
29. H. Seo, A. Posadas, A.A. Demkov, Strain-driven spin-state transition and superexchange interaction in $LaCoO_3$: *ab* initio study. Phys. Rev. B **86**, 014430 (2012)
30. T. Tatsuno, E. Mizoguchi, J. Nasu, M. Naka, S. Ishihara, Magnetic Field Effects in a Correlated Electron System with Spin-State Degree of Freedom- Implications for an Excitonic Insulator. J. Phys. Soc. Jpn. **85**, 083706 (2016)
31. J. Fernández Afonso and J. Kuneš, "Excitonic magnetism in d^6 perovskites" Phys. Rev. B 95, 115131 (2017)
32. J. Gazquez, W. Luo, M.P. Oxley, M. Prange, M.A. Torija, M. Sharma, C. Leighton, S.T. Pantelides, S.J. Pennycook, M. Varela, Atomic-resolution Imaging of Spin-State Superlattices in Nanopockets within Cobaltite Thin Films. Nano Lett. **11**, 973 (2011)
33. Y. Murakami et al., Resonant X-Ray Scattering from Orbital Ordering in $LaMnO_3$. Phys. Rev. Lett. **81**, 582 (1998)
34. V.E. Dmitrienko, Forbidden Reflections due to Anisotropic X-ray Susceptibility of Crystals. Acta Crystallogr. Sect. A **39**, 29 (1983)
35. S. Yamaguchi, Y. Okimoto, Y. Tokura, Local lattice distortion during the spin-state transition in $LaCoO_3$. Phys. Rev. B **55**, R8666 (1997)
36. H. Kruidhof, H.J.M. Bouwmeester, R. H. E. v. Doorn and A.J. Burggraaf, Influence of order-disorder transitions on oxygen permeability through selected nonstoichiometric perovskite-type oxides. Solid State Ionics 63–65 (1993) 816–822
37. J. Fujioka, Y. Yamasaki, A. Doi, H. Nakao, R. Kumai, Y. Murakami, M. Nakamura, M. Kawasaki, T. Arima, Y. Tokura, Strain-sensitive spin-state ordering in thin films of perovskite $LaCoO_3$. Phys. Rev. B **92**, 195115 (2015)
38. A. Kotani, S. Shin, Resonant inelastic x-ray scattering spectra for electrons in solids. Rev. Mod. Phys. **73**, 203 (2001)
39. S.G. Chiuzbăian, G. Ghiringhelli, C. Dallera, M. Grioni, P. Amann, X. Wang, L. Braicovich, L. Patthey, Localized electronic Excitations in NiO Studied with Resonant Inelastic X-Ray Scattering at the Ni M Threshold: Evidence of Spin Flip. Phys. Rev. Lett. **95**, 197402 (2005)
40. L.J.P. Ament, M. van Veenendaal, T.P. Devereaux, J.P. Hill, J. van den Brink, Resonant inelastic x-ray scattering studies of elementary excitations. Rev. Mod. Phys. **83**, 705 (2011)
41. Y. Yokoyama, Y. Yamasaki, M. Taguchi, Y. Hirata, K. Takubo, J. Miyawaki, Y. Harada, D. Asakura, J. Fujioka, M. Nakamura, H. Daimon, M. Kawasaki, Y. Tokura, H. Wadati, Tensile-strain-dependent Spin States in Epitaxial $LaCoO_3$ Thin Films. Phys. Rev. Lett. **120**, 206402 (2018)
42. Y. Harada, M. Kobayashi, H. Niwa, Y. Senba, H. Ohashi, T. Tokushima, Y. Horikawa, S. Shin, M. Oshima, Ultrahigh resolution soft x-ray emission spectrometer at BL07LSU in SPring-8. Rev. Sci. Instrum. **83**, 013116 (2012)
43. Y. Yamasaki, J. Fujioka, H. Nakao, J. Okamoto, T. Sudayama, Y. Murakami, M. Nakamura, M. Kawasaki, T. Arima, Y. Tokura, Surface ordering of orbitals at a higher temperature in $LaCoO_3$ thin Film. J. Phys. Soc. Jpn. **85**, 023704 (2016)
44. Y. Wakabayashi, M. Upton, S. Grenier, J. Hill, C. Nelson, J.-W. Kim, P. Ryan, A. Goldman, H. Zheng, J. Mitchell, Surface effects on the orbital order in the single-layered manganite $La_{0.5}Sr_{1.5}MnO_4$. Nat. Mater. **6**, 972 (2007)
45. S.B. Wilkins, X. Liu, Y. Wakabayashi, J.-W. Kim, P.J. Ryan, H. Zheng, J.F. Mitchell, J.P. Hill, Surface melting of electronic order in $La_{0.5}Sr_{1.5}MnO_4$. Phys. Rev. B **84**, 165103 (2011)

Chapter 6
Spin Transition in $BiCoO_3$ Correlated with Large Polar Distortion and Its Applications

Kengo Oka and Masaki Azuma

Abstract Perovskite oxide $BiCoO_3$ belongs to the class of compounds that show the spin transition of the Co ion. The spin state of Co^{3+} in $BiCoO_3$ is closely correlated with its crystal structure. The application of high pressure induces structural transition accompanied by spin transition. This chapter describes the correlation between the large structural distortion and the spin state of Co^{3+} in $BiCoO_3$. In addition, negative thermal expansion, which is observed in the $BiCoO_3$-$BaTiO_3$ system, and polarization rotation in the $BiCo_{1-x}Fe_xO_3$ solid solution system are introduced.

6.1 Synthesis and the Crystal Structure of $BiCoO_3$

6.1.1 Synthesis of $BiCoO_3$

Because of the stereochemical effect of 6 s^2 lone pair electrons, Bi^{3+}-containing oxides are a class of ferroelectric materials in which non-centrosymmetric structures induce spontaneous polarization in their lattice [1]. Among them, perovskite oxides containing Bi and 3d transition metal ions are attracting considerable attention as multiferroic materials in which ferroelectric and (anti)ferromagnetic properties coexist. Various $BiMO_3$ (M = Sc, Cr, Mn, Fe, Co, and Ni) perovskites have been reported and studied [2–7]. The majority of these perovskites (e.g., $BiScO_3$, $BiCrO_3$, $BiMnO_3$, $BiCoO_3$, and $BiNiO_3$) requires high-pressure (HP) conditions (4–8 GPa) for the synthesis, except for $BiFeO_3$, which can be synthesized under ambient pressure.

K. Oka (✉)
Department of Applied Chemistry, Faculty of Science and Engineering, Kindai University, Higashiosaka, Osaka 577-8502, Japan
e-mail: koka@apch.kindai.ac.jp

M. Azuma
Laboratory for Materials and Structures, Tokyo Institute of Technology, Yokohama 226-8503, Japan

Y. Okimoto et al. (eds.), *Spin-Crossover Cobaltite*, Springer Series in Materials Science 305, https://doi.org/10.1007/978-981-15-7929-5_6

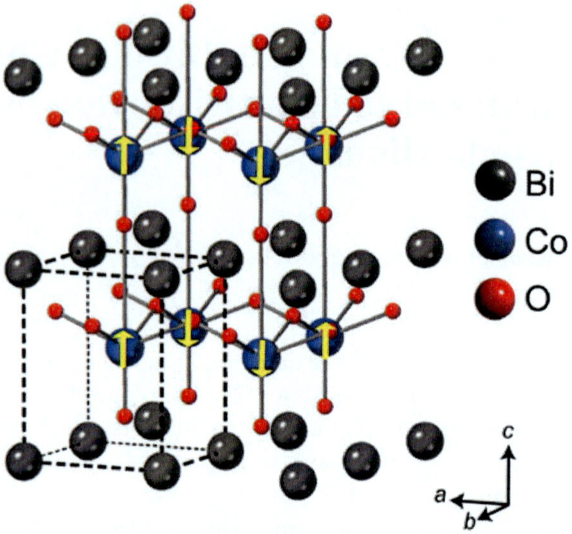

Fig. 6.1 Crystal structure of $BiCoO_3$ with the dashed line indicating the unit cell. The arrows indicate the *C*-type ordering of the magnetic moment below $T_N = 470$ K

$BiCoO_3$ was first synthesized by Belik et al. under the HP condition of 6 GPa and 1243 K [6]. Their structural analysis revealed that $BiCoO_3$ adopts the tetragonal perovskite structure with *P*4*mm* symmetry (Fig. 6.1). This crystal structure is isostructural with $PbTiO_3$, which is known as the mother compound of commercial piezoelectric material $PbZr_xTi_{1-x}O_3$ (PZT). $BiCoO_3$ exhibits spontaneous polarization in the crystal structure because *P*4*mm* space group is non-centrosymmetric. Belik et al. have also studied the magnetic properties and determined that $BiCoO_3$ is an antiferromagnet with the Néel temperature of $T_N = 470$ K. Their neutron powder diffraction study at 5 K indicates the presence of long-range antiferromagnetic ordering, which forms a *C*-type antiferromagnetic structure, as shown in Fig. 6.1. The refinement of the magnetic structure for $BiCoO_3$ provided a magnetic moment of 3.24(2) μ_B at 5 K and 2.93(2) μ_B at 300 K for the Co ion, which confirmed the high-spin state of Co^{3+} (HS: $S = 2$) rather than the low-spin (LS: $S = 0$) or intermediate spin (IS: $S = 1$) state.

6.1.2 Large Tetragonal Distortion Stabilized by the High-Spin State of Co^{3+}

The spin state of Co^{3+} in $BiCoO_3$ is closely correlated with its distorted crystal structure. The ratio of lattice parameters *c* and *a* (*c*/*a*) indicates the degree of structural distortion of tetragonal perovskites, and $BiCoO_3$ exhibits the large tetragonal distortion of $c/a = 1.27$, which is much larger than that of $PbTiO_3$ ($c/a = 1.06$) [8].

The origin of large tetragonal distortion was proposed by studying $PbVO_3$. Perovskite $PbVO_3$ is isostructural with $BiCoO_3$ and $PbTiO_3$ and also exhibits a large tetragonal distortion of $c/a = 1.23$ [9, 10]. The study of magnetic properties performed on the single crystalline $PbVO_3$ sample revealed the presence of low-dimensional magnetism, which resulted from the orbital order of the $3d^1$ electron of V^{4+} [11]. The mechanism of stabilizing large tetragonal distortion in $PbVO_3$ can be interpreted as follows. In the pyramidal coordination present in tetragonal perovskites, three t_{2g} orbitals increase its degeneracy into the non-degenerate d_{xy} orbital at the lower level and two degenerate d_{xz} and d_{yz} orbitals at the higher level. Thus, one d electron occupies a non-degenerate d_{xy} orbital, which results in the orbital order (Fig. 6.2) . Thus, low-dimensional magnetism is present in $PbVO_3$. In the pyramidal V–O configuration, one of the V–O_{apex} bonds is significantly shorter than the V–$O_{in\text{-}plane}$ bonds; therefore, the energy of the d_{xy} orbital is lowered, i.e., the tetragonal distortion is enhanced in the d^1 system for the energy gain. The same scenario can be seen in $BiCoO_3$. When Co^{3+} in $BiCoO_3$ adopts a high-spin electronic configuration, five electrons are distributed to each of the five d orbitals, and the last electron goes into the non-degenerate d_{xy} orbital as in the d^1 system, as shown in Fig. 6.3. Thus, the high-spin state of Co^{3+} stabilizes the crystal structure of tetragonal $BiCoO_3$ with large tetragonal distortion.

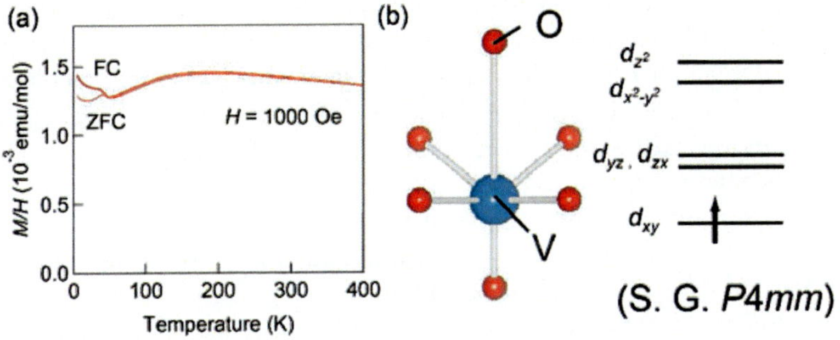

Fig. 6.2 **a** Temperature dependence of the magnetic susceptibility of $PbVO_3$. The broad maximum, which is characteristic of a low-dimensional magnet, is present at 180 K. **b** Coordination state around V^{4+} in $PbVO_3$ with the energy level diagram of the 3d orbital. Reprinted with permission from [11]. Copyright 2008 American Chemical Society

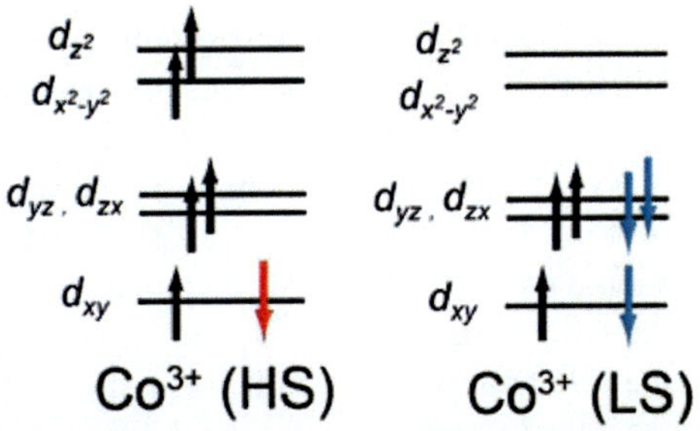

Fig. 6.3 Schematic drawings of the 3d orbital of Co^{3+} in tetragonal $BiCoO_3$ for both high-spin (HS) and low-spin (LS) states

6.2 Spin Transition in $BiCoO_3$ with a Large Volume Collapse

6.2.1 Pressure-Induced Structural Transition of $BiCoO_3$

The HS state of Co^{3+} acts to stabilize the large tetragonal distortion of $BiCoO_3$. In general, ferroelectric materials change into the paraelectric phase at high temperatures above the Curie temperature (T_C). For example, $PbTiO_3$ changes from a polar tetragonal structure (*P*4*mm*) to the centrosymmetric cubic perovskite structure (*Pm*-3 *m*) at $T_C = 763$ K [12]. Therefore, we can expect the presence of the spin state change that is accompanied by the structural transition. However, $BiCoO_3$ decomposes above 720 K without showing a structural transition to a paraelectric phase. According to an empirical relation between T_C and the spontaneous polarization P_S (μC/cm^2), $T_C = (0.303 \pm 0.018)P_S^2$, [13] the T_C of $BiCoO_3$ is expected to be ~4500 K. Thus, the observation of the temperature-induced structural transition of $BiCoO_3$ without sample decomposition is almost impossible due to the large distortion.

An alternative way to induce structural transition to the paraelectric phase is to apply HP. $PbTiO_3$ exhibits a sequence of phase transitions from tetragonal to monoclinic to another monoclinic and to rhombohedral phases with an increase in the pressure [14, 15]. $PbVO_3$ also undergoes structural transition to the cubic phase (*Pm*-3 *m*) above 3 GPa accompanied by the insulator to metal transition, which results from the degeneration of t_{2g} orbitals [16, 17].

The pressure-induced structural transition of $BiCoO_3$ has been observed by both synchrotron X-ray (SXRD) and time-of-flight neutron powder diffractions (NPD) under HP conditions (Fig. 6.4) [18]. The HP phase of $BiCoO_3$ appears at 2.4 GPa, and the structural transition is complete above 3 GPa owing to the ambient pressure (AP) and HP phase-coexisting state. This behavior demonstrates the first-order nature of the structural transition, which is similar to that of $PbVO_3$. The structural analysis of the NPD pattern at 5.8 GPa revealed that the HP phase of $BiCoO_3$ is $GdFeO_3$-type orthorhombic (*Pbnm*) perovskite (Fig. 6.4). The refined crystallographic parameters for the AP and HP phases are summarized in Table 6.1. The unit cell of the HP phase expands to the $\sqrt{2}\, a \times \sqrt{2}\, a \times 2c$ superstructure from the AP phase. Because the space group *Pbnm* is centrosymmetric and non-polar, this structural transition can be classified as the ferroelectric-paraelectric transition. It is also interesting to point out the similarities between the HP phases of $BiMO_3$ (M = Sc, [19] Cr, [19] Fe, [20] Mn[19] and Ni[21]), which are all known to adopt the same $GdFeO_3$-type structure.

Next, let us compare the crystal structure of the AP and HP phases of $BiCoO_3$. It should be noted that a large volume collapse reaching 15 vol.% is present between these two phases. By comparing the lattice parameters of both phases, it is determined that the large volume collapse is due to the shrinkage of the c axis, which results from the loss of large tetragonal distortion (Fig. 6.5). Although $BiCoO_3$ decomposes without showing structural transition under AP, the temperature-induced structural

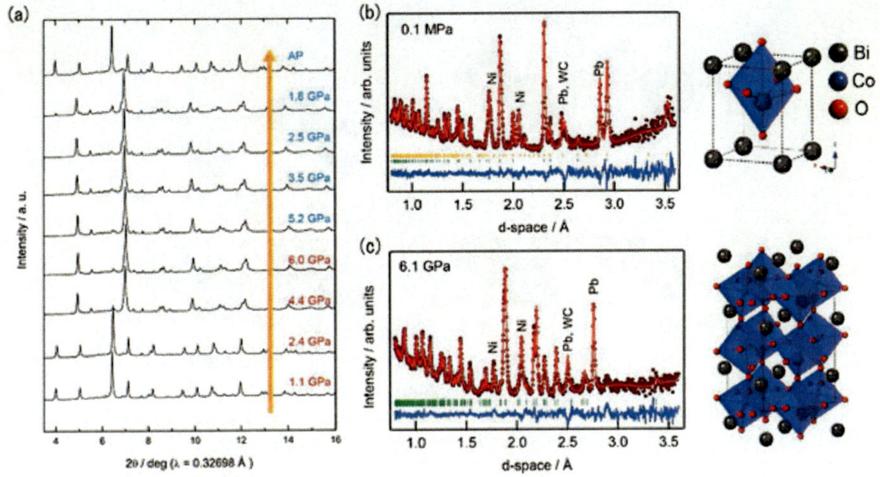

Fig. 6.4 **a** Synchrotron X-ray ($\lambda = 0.39328$ Å) diffraction patterns of $BiCoO_3$ at room temperature collected at various pressures during compression (red letters) and decompression (blue letters) processes. **b** Time-of-flight neutron powder diffraction patterns and the results of Rietveld refinement for $BiCoO_3$ at 300 K. The patterns were collected at 0.1 MPa (upper panel) and 6.1 GPa (bottom panel). The upper and lower tick marks for the AP phase represent the crystal diffraction and magnetic diffraction, respectively. (The contributions of Pb as a pressure marker and Ni and tungsten carbide (WC) from the apparatus are also fitted.) The illustration of the crystal structures of $BiCoO_3$ obtained by the refined crystallographic parameters for both the AP and HP phases is shown on the right side of the neutron diffraction patterns. Reprinted with permission from [18]. Copyright 2010 American Chemical Society

Table 6.1 Refined crystallographic parameters from the NPD patterns of the AP and HP phases of $BiCoO_3$

Atom	Site	x	y	z	U_{iso} ($Å^2$)
AP phase (0.1 MPa) Space group *P*4mm, $a = 3.7310(2)$ Å, $c = 4.7247(5)$ Å, m_z (Co) $= 3.08(9)$ μ_B, $Z = 1$					
Bi	1a	0	0	0	0.0015(7)
Co	1b	0.5	0.5	0.5699(22)	0.0015(7)
O1	1b	0.5	0.5	0.2001(13)	0.0114(10)
O2	2c	0.5	0	0.7266(9)	0.0042(8)
HP phase (6.1 GPa) Space group *P*bnm, $a = 5.2963(7)$ Å, $b = 5.3936(6)$ Å, $c = 7.5469(11)$ Å, $Z = 4$					
Bi	4c	0.9960(14)	0.0417(8)	0.25	0.0004(6)
Co	4a	0	0.5	0	0.0004(6)
O1	4c	0.0777(15)	0.4844(12)	0.25	0.0038(13)
O2	8d	0.7067(10)	0.2906(9)	0.0404(7)	0.0054(10)

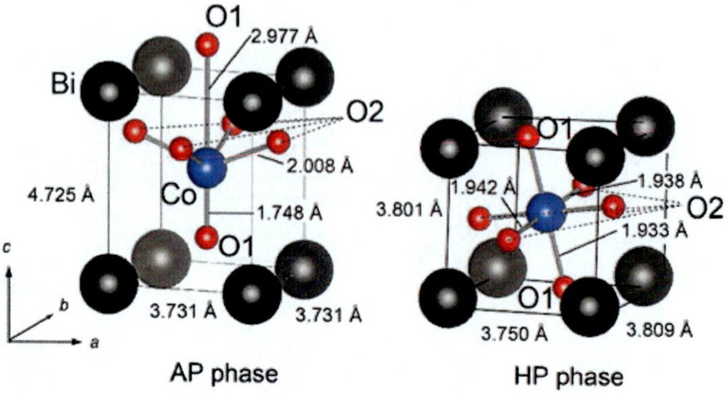

Fig. 6.5 Coordination states around the Co ion in the AP and HP phases of $BiCoO_3$. Reprinted with permission from [18]. Copyright 2010 American Chemical Society

transition can be observed under moderate pressure via the region with two coexisting phases. Figure 6.6 shows the pressure–temperature (P–T) phase diagram of $BiCoO_3$ and the temperature dependence of the cell volume per formula unit with the weighted average volume of the AP phase at 2.4 GPa. Under the moderate pressure of 2.4 GPa, the temperature-induced structural transition from the AP phase to the HP phase is observed. As shown in Fig. 6.6b, the average unit cell volume shows an unusual behavior; it linearly decreases upon heating. The phenomenon that the volume contracts upon heating is the so-called negative thermal expansion (NTE). $BiCoO_3$ is a promising candidate of the NTE material, i.e., $PbVO_3$-based materials [22–24]. Indeed, the suppression of the tetragonal distortion in the $BiCoO_3$–$BaTiO_3$ system has been reported [25]. The tetragonality, c/a, reduces upon an increase in the $BaTiO_3$ content, and the structural transition from the tetragonal to cubic phase

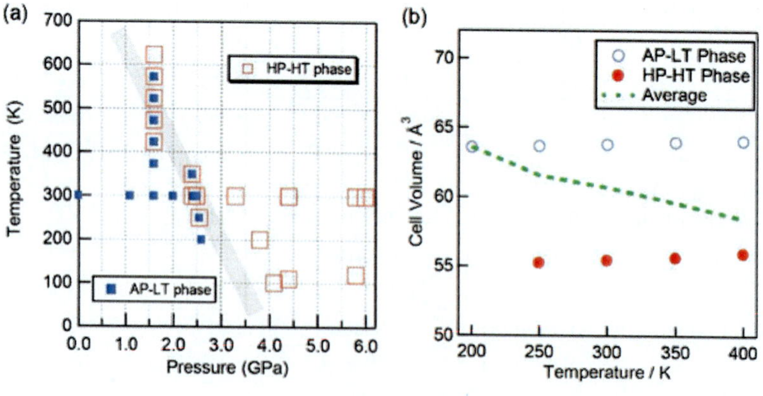

Fig. 6.6 **a** Pressure–temperature (P–T) phase diagram of $BiCoO_3$. The shaded region indicates the phase boundary. **b** Temperature dependence of the cell volume per formula unit and the average unit cell volume at 2.4 GPa. Reprinted with permission from [18]. Copyright 2010 American Chemical Society

occurs under AP. At the composition of 0.69 of $BiCoO_3$–$0.31BaTiO_3$, the *c*/*a* ratio is reduced to 1.15, and an NTE in the temperature range of 625–780 K with the total volume shrinkage of 4.4% has been observed.

6.2.2 Spin Transition Accompanied by the Structural Transition

It is important to determine how the physical properties change between the AP and HP phases of $BiCoO_3$. A drop in electrical resistivity and the structural transition simultaneously occur, as shown in Fig. 6.7a. This behavior indicates that the splitting of *d* orbitals into b_{2g} (d_{xy}), doubly degenerate e_g (d_{xz} and d_{yz}), a_{1g} (d_{z2}), and b_{1g} (d_{x2-y2}) levels in the pyramidal coordination changes to triply degenerate t_{2g} (d_{xy}, d_{xz}, d_{yz}) and doubly degenerate e_g (d_{z2} and d_{x2-y2}) in the octahedral pseudo-cubic coordination.

The LS state of Co^{3+} is expected for the HP phase because three 3*d* orbitals degenerate into t_{2g}. Indeed, the Rietveld refinement performed on the NPD pattern of the HP phase has been successfully converged without a magnetic component, which indicates the negligible magnetic moment of the Co^{3+}. The definitive evidence of the spin state change has been provided by Co *K*β X-ray emission spectroscopy (XES). Figure 6.7b shows the Co *K*β XES spectra of the AP and HP phases. Two components, $K\beta_{1,3}$ and *K*β', are present in the spectra. It should be noted that $K\beta'$, which is characteristic of the high-spin state, is less pronounced in the spectrum of the HP phase, which indicates the spin state change of Co^{3+}, as reported for $LaCoO_3$ [26–28]. The integration of absolute value of the difference (IAD) spectra allows to empirically estimate ΔS; IAD = 0.049 for D*S* = 1, 0.084 for D*S* = 3/2, and 0.12 for D*S* = 2 [26, 29, 30]. The integration of the difference $\sum |I_{AP} - I_{HP}|$ in the energy range of 7615–7670 eV, where the suffixes AP and HP indicate the normalized intensity for the AP and HP phases, gives the IAD value of 0.041 for BiCoO-$_3$. This magnitude corresponds to $\Delta S = 1$, i.e., the spin state change from the high-spin ($S = 2$) to the intermediate spin ($S = 1$) state for Co^{3+}, as shown in Fig. 6.7(c). For Co^{3+} of the HP phase, there is a possibility to show the LS state at low temperatures, as observed for $LaCoO_3$, because these XES spectra were obtained at room temperature.

6.3 Piezoelectric Property in $BiCoO_3$-Based Materials

The materials exhibiting spontaneous electronic polarization (e.g., $BaTiO_3$ and $PbTiO_3$) are expected to show piezoelectric properties. In addition, $BiCoO_3$ is a promising piezoelectric material because of its large tetragonal distortion. Analogous to $PbZr_xTi_{1-x}O_3$ (PZT), a well-known piezoelectric material, the solid solution

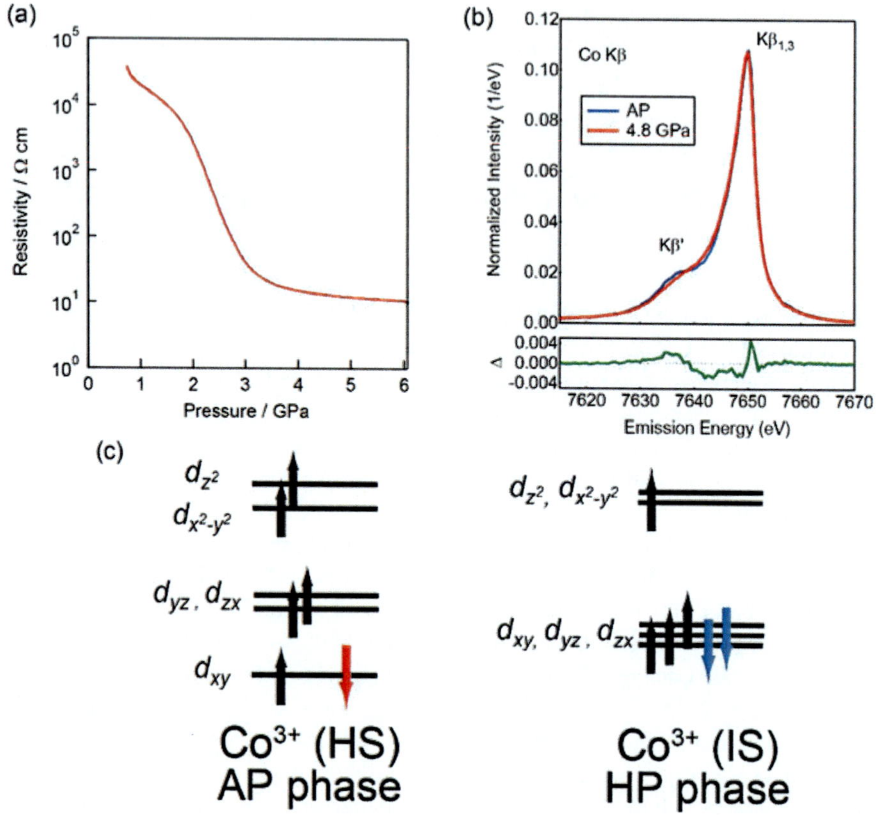

Fig. 6.7 **a** Pressure dependence of the resistivity of $BiCoO_3$ at room temperature. **b** Co $K\beta$ X-ray emission spectra (XES) of the AP (0.1 MPa) and HP (4.8 GPa) phases of $BiCoO_3$. Both spectra were normalized to have integrated intensity = 1.0 in the energy range of 7615–7670 eV. Δ is the difference of normalized intensity, $I_{AP} - I_{HP}$. **c** Schematic energy diagrams of the 3d orbital for the AP and HP phases of Co^{3+} in $BiCoO_3$. Reprinted with permission from [18]. Copyright 2010 American Chemical Society

of $BiCoO_3$ and rhombohedral perovskite $BiFeO_3$ has been studied as a piezoelectric material.

Figure 6.8 shows the phase diagram of the $BiCoO_3$–$BiFeO_3$ system. At the phase boundary of the tetragonal (Co-rich) and rhombohedral (Fe-rich) phases, which is the so-called morphotropic phase boundary (MPB), a monoclinic structure appears and gradually changes to the tetragonal phase at high temperatures [31]. This monoclinic perovskite phase at MPB is very important for a piezoelectric material, because PZT shows the highest piezoelectric response for the monoclinic phase [32]. For the tetragonal and rhombohedral perovskite phases, the polarization vectors are fixed at the (001) and (111) direction of cubic perovskite, respectively. However, in the monoclinic phase, lowered symmetry allows the rotation of the polarization, as shown

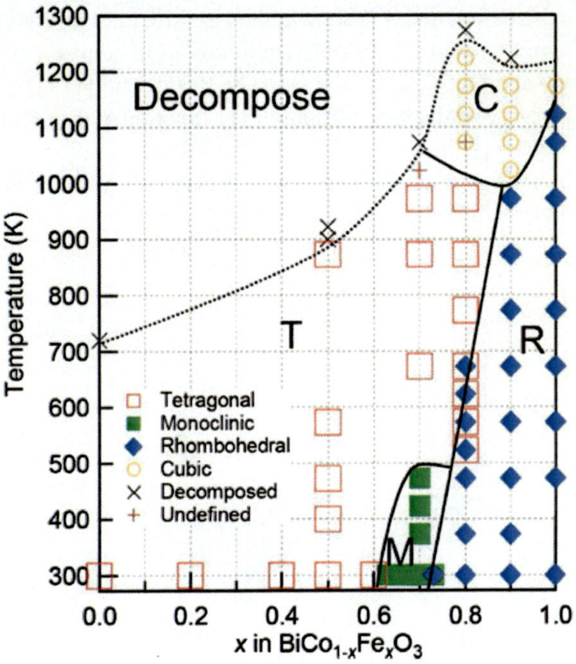

Fig. 6.8 Composition–temperature phase diagram for the $BiCo_{1-x}Fe_xO_3$ system. T, R, M, and C denote the tetragonal, rhombohedral, monoclinic, and cubic phases, respectively. (the data were taken from Refs. [31 and 34].)

in Fig. 6.9. The superior piezoelectric property of PZT at MPB has been interpreted by considering polarization rotation that is induced by the applied electric field [33].

Figure 6.10a shows the monoclinic structure of $BiCo_{1-x}Fe_xO_3$ ($x = 0.3$) with a $\sqrt{2}\,a \times \sqrt{2}\,a \times a$ unit cell, where a is the lattice parameter for cubic perovskite. Accurate structural analysis revealed a non-centrosymmetric space group *Cm* for

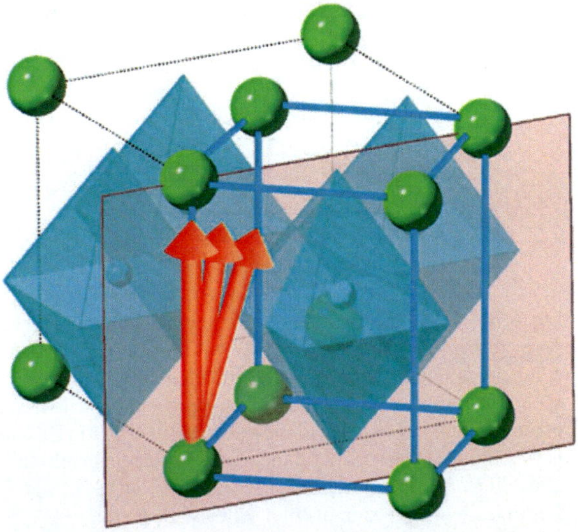

Fig. 6.9 Schematic illustration of the polarization rotation in the monoclinic perovskite phase. The dotted and blue lines denote the $\sqrt{2}\,a \times \sqrt{2}\,a \times a$ supercell for the monoclinic phase and the original perovskite unit cells, respectively. The polarization vectors are indicated by the arrows and can rotate within the $(110)_c$ plane (marked in pink)

the monoclinic phase, which is isostructural with that of PZT at MPB [34]. The *Cm* space group allows polarization rotation between the [001] and [111] directions of a pseudo-cubic cell (respectively, $[001]_c$ and $[111]_c$). The SXRD study allowed us to observe polarization rotation as a function of the temperature for the $BiCoO_3$–$BiFeO_3$ system (Fig. 6.10). The spontaneous polarization estimated by the point charge model shows a gradual change in the polarization vector between the monoclinic (Fe-rich, low temperature) to tetragonal (Co-rich, high temperature) phases, which indicates its potential for a piezoelectric material. Indeed, the piezoelectric property of the

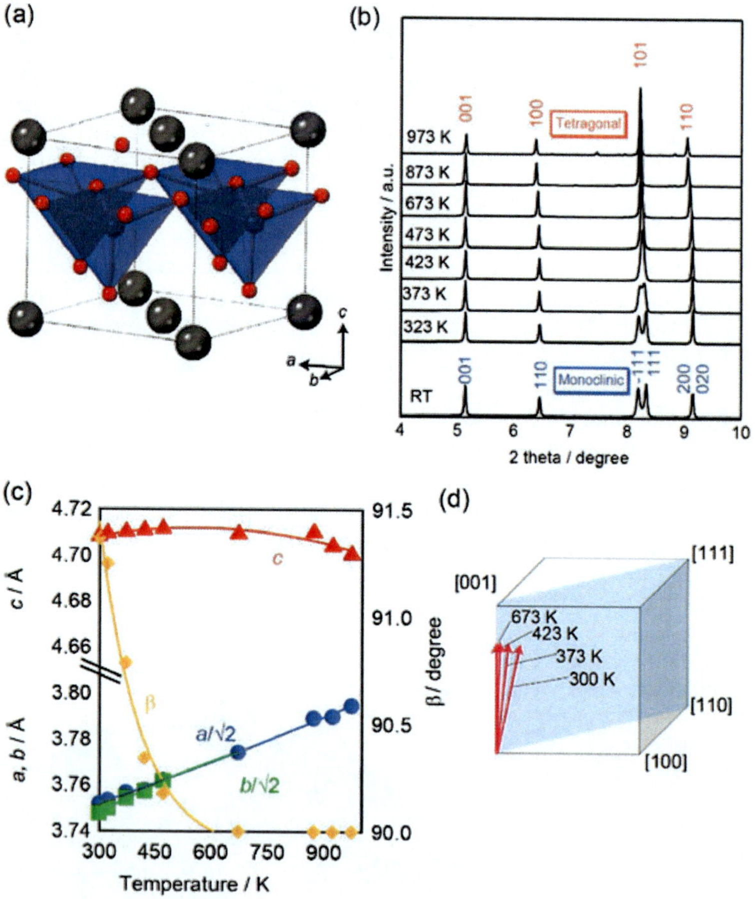

Fig. 6.10 **a** Crystal structure of the monoclinic phase of $BiCo_{1-x}Fe_xO_3$ (S.G. *Cm*). **b** Synchrotron X-ray diffraction patterns ($\lambda = 0.4229$ Å) for $BiCo_{0.70}Fe_{0.30}O_3$ collected at elevated temperatures. **c** Temperature dependence of the lattice parameters for $BiCo_{0.70}Fe_{0.30}O_3$. A decrease in β corresponds to the structural transition from the monoclinic to tetragonal phase. **d** Schematic illustration of temperature dependence of the polarization vector in $BiCo_{0.70}Fe_{0.30}O_3$. The indices [001], [100], [110], and [111] are based on the pseudo-cubic unit cell. The rotation of the polarization vector as a function of temperature is clearly shown. From [34] Copyright © 2000 by John Wiley Sons, Inc. Reprinted by permission of John Wiley & Sons, Inc

monoclinic phase has been studied for $BiCoO_3$–$BiFeO_3$ thin films, and piezoelectric responses are observed for the monoclinic phases [35].

6.4 Concluding Remarks

The structural property of $BiCoO_3$ is strongly correlated with the spin state of Co^{3+}. The non-centrosymmetric pyramidal coordination of Co^{3+} induced by the steric effect of Bi^{3+} can be stabilized by the HS configuration of Co^{3+} with $3d^6$ electrons, which leads to a large tetragonal distortion reaching $c/a = 1.27$. $BiCoO_3$ undergoes a structural transition from the tetragonal-ferroelectric to the orthorhombic-paraelectric phase under HP conditions. This structural transition is accompanied by the spin state change of Co^{3+} from the HS to IS/LS state. A large volume collapse of 10% simultaneously occurs owing to the elimination of the large tetragonal distortion. These exotic structural properties can be applied to NTE or piezoelectric materials by chemical modification. The tuning of the Co^{3+} spin state in $BiCoO_3$ is a promising approach for developing its potential applications.

References

1. A.A. Belik, J. Solid State Chem. **195**, 32–40 (2012)
2. A.A. Belik, S. Iikubo, K. Kodama, N. Igawa, S-i. Shamoto, M. Maie, T. Nagai, Y. Matsui, S.Y. Stefanovich, B.I. Lazoryak, E. Takayama-Muromachi, J. Am. Chem. Soc. **128**(3), 706–707 (2006)
3. S. Niitaka, M. Azuma, M. Takano, E. Nishibori, M. Takata, M. Sakata, Solid State Ionics **172**(1), 557–559 (2004)
4. T. Atou, H. Chiba, K. Ohoyama, Y. Yamaguchi, Y. Syono, J. Solid State Chem. **145**(2), 639 (1999)
5. C. Michel, J.-M. Moreau, G.D. Achenbach, R. Gerson, W.J. James, Solid State Commu. **7**(9), 701–704 (1969)
6. A.A. Belik, S. Iikubo, K. Kodama, N. Igawa, S-i. Shamoto, S. Niitaka, M. Azuma, Y. Shimakawa, M. Takano, F. Izumi, E. Takayama-Muromachi, Chem. Mater. **18**(3), 798–803 (2006)
7. S. Ishiwata, M. Azuma, M. Takano, E. Nishibori, M. Takata, M. Sakata, K. Kato, J. Mater. Chem. **12**(12), 3733–3737 (2002)
8. Y. Kuroiwa, S. Aoyagi, A. Sawada, J. Harada, E. Nishibori, M. Takata, M. Sakata, Phys. Rev. Lett. **87**(21), 217601 (2001)
9. A.A. Belik, M. Azuma, T. Saito, Y. Shimakawa, M. Takano, Chem. Mater. **17**(2), 269–273 (2005)
10. R.V. Shpanchenko, V.V. Chernaya, A.A. Tsirlin, P.S. Chizhov, D.E. Sklovsky, E.V. Antipov, E.P. Khlybov, V. Pomjakushin, A.M. Balagurov, J.E. Medvedeva, E.E. Kaul, C. Geibel, Chem. Mater. **16**(17), 3267–3273 (2004)
11. K. Oka, I. Yamada, M. Azuma, S. Takeshita, K.H. Satoh, A. Koda, R. Kadono, M. Takano, Y. Shimakawa, Inorg. Chem. **47**(16), 7355–7359 (2008)
12. G. Shirane, R. Pepinsky, B.C. Frazer, Acta. Crysta. **9**(2), 131–140 (1956)
13. S.C. Abrahams, S.K. Kurtz, P.B. Jamieson, Phys. Rev. **172**(2), 551–553 (1968)

14. M. Ahart, M. Somayazulu, R.E. Cohen, P. Ganesh, P. Dera, H-k. Mao, R.J. Hemley, Y. Ren, P. Liermann, Z. Wu, Nature **451**, 545–548 (2008)
15. P.E. Janolin, P. Bouvier, J. Kreisel, P.A. Thomas, I.A. Kornev, L. Bellaiche, W. Crichton, M. Hanfland, B. Dkhil, Phys. Rev. Lett. **101**(23), 237601 (2008)
16. A.A. Belik, T. Yamauchi, H. Ueda, Y. Ueda, H. Yusa, N. Hirao, M. Azuma, J. Phys. Soc. Jpn. **83**(7), 074711 (2014)
17. K. Oka, T. Yamauchi, S. Kanungo, T. Shimazu, K. Oh-ishi, Y. Uwatoko, M. Azuma, T. Saha-Dasgupta, J. Phys. Soc. Jpn. **87**(2), 024801 (2018)
18. K. Oka, M. Azuma, W-t. Chen, H. Yusa, A.A. Belik, E. Takayama-Muromachi, M. Mizumaki, N. Ishimatsu, N. Hiraoka, M. Tsujimoto, M.G. Tuchker, J.P. Attfield, Y. Shimakawa, J. Am. Chem. Soc. **132**(27), 9438–9443 (2010)
19. A.A. Belik, H. Yusa, N. Hirao, Y. Ohishi, E. Takayama-Muromachi, Inorg. Chem. **48**(3), 1000–1004 (2009)
20. R. Haumont, P. Bouvier, A. Pashkin, K. Rabia, S. Frank, B. Dkhil, W.A. Crichton, C.A. Kuntscher, J. Kreisel, Phys. Rev. B **79**(18), 184110 (2009)
21. M. Azuma, S. Carlsson, J. Rodgers, M.G. Tucker, M. Tsujimoto, S. Ishiwata, S. Isoda, Y. Shimakawa, M. Takano, J.P. Attfield, J. Am. Chem. Soc. **129**(46), 14433–14436 (2007)
22. H. Yamamoto, T. Imai, Y. Sakai, M. Azuma, Angew. Chem. Int. Ed. **57**(27), 8170–8173 (2018)
23. T. Ogata, K. Oka, M. Azuma, Appl. Phys. Express **12**(2), 023005 (2019)
24. T. Ogata, Y. Sakai, H. Yamamoto, S. Patel, P. Keil, J. Koruza, S. Kawaguchi, Z. Pan, T. Nishikubo, J. Rödel, M. Azuma, Chem. Mater. **31**(4), 1352–1358 (2019)
25. Z. Pan, X. Jiang, T. Nishikubo, Y. Sakai, H. Ishizaki, K. Oka, Z. Lin, M. Azuma, Chem. Mater. **31**(16), 6187–6192 (2019)
26. G. Vankó, J.-P. Rueff, A. Mattila, Z. Németh, A. Shukla, Phys. Rev. B **73**(2), 024424 (2006)
27. D.P. Kozlenko, N.O. Golosova, Z. Jirák, L.S. Dubrovinsky, B.N. Savenko, M.G. Tucker, Y. Le Godec, V.P. Glazkov, Phys. Rev. B **75**(6), 064422 (2007)
28. R. Lengsdorf, J.P. Rueff, G. Vankó, T. Lorenz, L.H. Tjeng, M.M. Abd-Elmeguid, Phys. Rev. B **75**(18), 180401 (2007)
29. G. Vankó, T. Neisius, G. Molnár, F. Renz, S. Kárpáti, A. Shukla, F.M.F. de Groot, J. Phys. Chem. B. **110**(24), 11647–11653 (2006)
30. J.P. Rueff, A. Shukla, A. Kaprolat, M. Krisch, M. Lorenzen, F. Sette, R. Verbeni, Phys. Rev. B **63**(13), 132409 (2001)
31. M. Azuma, S. Niitaka, N. Hayashi, K. Oka, M. Takano, H. Funakubo, Y. Shimakawa, Jpn. J. Appl. Phys. **47**(9), 7579–7581 (2008)
32. B. Noheda, J.A. Gonzalo, L.E. Cross, R. Guo, S.E. Park, D.E. Cox, G. Shirane, Phys. Rev. B **61**(13), 8687–8695 (2000)
33. D. Vanderbilt, M.H. Cohen, Phys. Rev. B **63**(9), 094108 (2001)
34. K. Oka, T. Koyama, T. Ozaaki, S. Mori, Y. Shimakawa, M. Azuma, Angew. Chem. Int. Ed. **51**(32), 7977–7980 (2012)
35. H. Hojo, K. Oka, K. Shimizu, H. Yamamoto, R. Kawabe, M. Azuma, Adv. Mater. **30**(33), 1705665 (2018)

Chapter 7
Thermoelectric Properties of Cobalt Oxides and Other Doped Mott Insulators

Wataru Koshibae

Abstract One of the most important topics of cobalt oxides is the large thermopower (absolute Seebeck coefficient), which has been attracted attention due to their potential application as thermoelectric conversion material. In this chapter, the basics of thermoelectrics and thermoelectric response in cobalt oxides and related transition metal oxides are introduced. Thermopower is corresponding to the carried entropy by the electric current. In the strongly correlated electron systems, the spin and orbital degrees of freedom contribute to the entropy flow. The role of spin and orbital degrees of freedom on the thermoelectric effect is theoretically discussed.

7.1 Introduction

Terasaki et al. [1] have found the large thermopower and the low resistivity in the cobalt oxide Na_2CoO_4. After the discovery, a number of experimental and theoretical studies [2–32] on the oxide thermoelectrics have been developed.

In the cobalt oxide Na_2CoO_4, it has been observed that the electronic transport occurs on the $3dt_{2g}$ manifold of the cobalt ion(s). The related cobalt oxides [3–5, 9, 10, 14, 23, 31, 32] also show the large thermopower. These large thermopowers often occur above room temperatures. In particular, it has been reported [5, 8] that the figure of merit of the thermoelectrics, $ZT = Q^2/(\rho\kappa)$ (Q: thermopower, ρ: resistivity, κ: thermal conductivity) can be $ZT > 1$ (a criterion for good thermoelectric material) above room temperatures for some of the cobalt oxides.

In this chapter, we discuss the thermopower at high temperatures in the strongly correlated electron systems. To this end, first, let us review the basics of thermoelectric response in solid [17, 33–35], and clarify the physical meaning of the thermopower, in the Sects. 7.2–7.3. In Sects. 7.4 and 7.6, the formula for the high-temperature

W. Koshibae (✉)
RIKEN Center for Emergent Matter Science (CEMS), Wako, Japan
e-mail: wataru@riken.jp

Y. Okimoto et al. (eds.), *Spin-Crossover Cobaltite*, Springer Series in Materials Science 305, https://doi.org/10.1007/978-981-15-7929-5_7

thermopower is derived. Based on the formula, in Sect. 7.7, we discuss the thermoelectric response of the cobalt oxide and that of vanadium oxide as a related strongly correlated electron system. Finally, we discuss the oxide thermoelectrics in Sect. 7.8.

7.2 Thermoelectric Effect

In the conductors, we often see the cross effect between the electricity and heat, e.g., heat flows along with electric current and temperature gradient drives the electric current. The typical thermoelectric effect, Seebeck effect, is useful to measure the temperature of the sample, and in some cases, it is utilized for the electric power generation. Let us first discuss the Seebeck effect and the meaning of the measurements on voltmeter.

Figure 7.1a shows a thermocouple of the conductors A and B. The conductor B is disconnected, so that the electric current (direct current) is not flowing. The Seebeck effect describes that the voltage V_{AB} is induced by the (small) temperature difference $\Delta T(>0)$. Figure 7.1b is an experimental setup to measure the voltage V_{AB}: The terminal of the voltmeter denoted by "0" ("5") and an open end of the conductor B denoted by "1" ("4") are connected by the lead wire. The temperature at the open ends, "1" and "4", is denoted by T_s. The voltage V_{AB} reflects the absolute Seebeck coefficient or so-called thermopower of A and B. The thermopower Q is defined by the gradient of the electro-chemical potential induced by the temperature gradient,

$$-\frac{\partial \varphi}{\partial x}+\frac{1}{e}\frac{\partial \mu}{\partial x}=Q\frac{\partial T}{\partial x}, \tag{7.1}$$

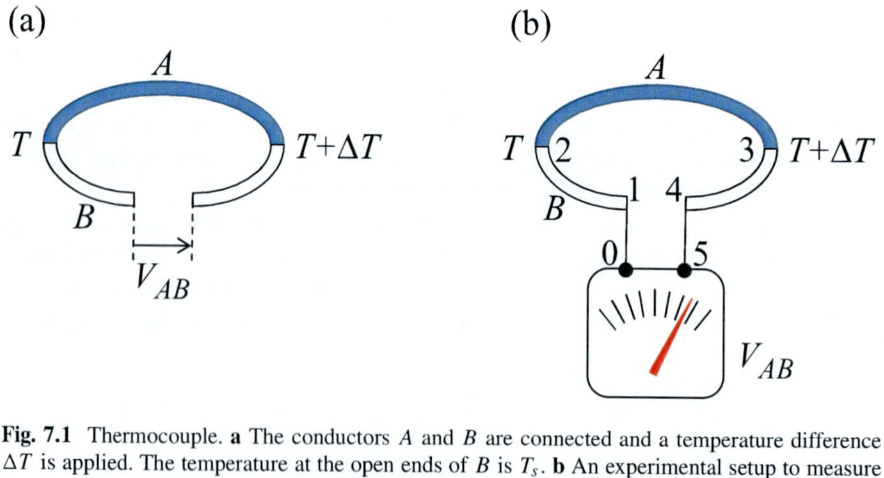

Fig. 7.1 Thermocouple. **a** The conductors A and B are connected and a temperature difference ΔT is applied. The temperature at the open ends of B is T_s. **b** An experimental setup to measure the voltage V_{AB} of the thermocouple. An open end of B denoted by "1" ("4") and a terminal of the voltmeter denoted by "0" ("5") are connected by lead wire

where φ is the electro-static potential, μ is the chemical potential, and $e(>0)$ is electron charge. Let us examine the relation between the thermopower of A and B, and the voltage V_{AB}. On the conductor A, we find

$$\int_2^3 \left(-\frac{\partial \varphi_A}{\partial x} + \frac{1}{e}\frac{\partial \mu_A}{\partial x}\right) dx = \frac{1}{e}\left(\mu_A - e\varphi_A\right)_3 - \frac{1}{e}\left(\mu_A - e\varphi_A\right)_2 . \quad (7.2)$$

and

$$\int_2^3 Q_A \frac{\partial T}{\partial x} dx = \int_T^{T+\Delta T} Q_A dT = Q_A \Delta T, \quad (7.3)$$

where φ_A, μ_A and Q_A represent the electro-static potential, the chemical potential and thermopower of A, respectively, and x is taken along the conductor A. In the same way, we perform the integration along the circuit "0" → "1" → "2" → "3" → "4" → "5" in Fig. 7.1b. Under the condition where no electric current is flowing in the circuit, there is no difference of the electro-chemical potential between the conductors forming the junctions "1", "2", "3", and "4", e.g., $(\mu_A - e\varphi_A)_2 = (\mu_B - e\varphi_B)_2$ at the junction "2", where φ_B and μ_B represent the electro-static potential and the chemical potential of B, respectively. Therefore, we find

$$\begin{aligned} &\frac{1}{e}\left(\mu_l - e\varphi_l\right)_5 - \frac{1}{e}\left(\mu_l - e\varphi_l\right)_0 \\ &= \int_{T_0}^{T_1} Q_l dT + \int_{T_1}^{T_2} Q_B dT + \int_{T_2}^{T_3} Q_A dT + \int_{T_3}^{T_4} Q_B dT + \int_{T_4}^{T_5} Q_l dT, \end{aligned} \quad (7.4)$$

where φ_l, μ_l and Q_l are the electro-static potential, the chemical potential and thermopower of the lead wire, respectively, and Q_B is the thermopower of B. The temperatures at the junctions "0"~"5" are denoted by $T_0 \sim T_5$. By putting $T_0 = T_5$, $T_1 = T_4 (= T_s)$, $T_2 = T$ and $T_3 = T + \Delta T$, Eq. (7.4) for small ΔT gives

$$V_{AB} = (\varphi_l)_5 - (\varphi_l)_0 = \int_T^{T+\Delta T} (Q_B - Q_A)\, dT = [Q_B(T) - Q_A(T)]\, \Delta T. \quad (7.5)$$

Using the reference conductor B which thermopower is known, we can measure the thermopower of the conductor A by the voltmeter. The superconductor defines *zero* of the thermopower and is useful for the reference conductor. The use of superconductivity as the reference conductor is limited for the measurements at low temperatures. In the normal-conducting environment, the thermopower of the typical reference materials, e.g., Pb, Pt, etc., is determined by the Thomson effect which is a thermoelectric effect different from the Seebeck effect. Next, let us discuss the thermoelectric effects beyond the Seebeck effect. By the consideration below, we find the low-temperature behavior of the thermopower.

7.3 Thermoelectric Effects and Laws of Thermodynamics

In the previous section, we discussed the thermoelectric effect without the electric current flow. In addition to the Seebeck effect, we find the other thermoelectric effects [34].

A heat emission/absorption occurs at the junction between the conductors when an electric current flows through the junction. This is called the Peltier effect. Figure 7.2a shows the Peltier effect of the thermocouple by the conductors A and B: The temperature is uniform throughout the thermocouple. The electric current I is flowing in the thermocouple. The heat emitted per unit time and per unit electric current I at the left junction between the conductors A and B defines the Peltier coefficient of the Π_{AB}. Obviously, it is found that $\Pi_{BA} = -\Pi_{AB}$. It is also found that $\Pi_{AA} = 0$, i.e., this thermoelectric effect does not occur in the uniform conductors without temperature gradient.

In the uniform conductors *with* temperature gradient shows thermoelectric effect called Thomson effect. Figure 7.2b explains the Thomson effect: The conductor has the temperature gradient ($\Delta T > 0$) and the electric current I is flowing from the low-temperature side to high-temperature side. In this setup, to maintain the temperature gradient, the heat (Δq) is absorbed by the conductor per unit time, unit current, and per temperature gradient and defines the Thomson coefficient τ, i.e.,

$$\Delta q = \tau(T)I\left(\frac{\partial T}{\partial x}\right)\Delta x = \tau(T)I\Delta T. \tag{7.6}$$

This is understood as the heat required to change the temperature of the electric current I, therefore Thomson described [34] this coefficient τ as the "specific heat of electricity". (Note that τ can be negative depending on the material.)

The thermoelectric effects discussed above, the Seebeck effect, the Peltier effect, and Thomson effect, are *not* independent ones, and Thomson [34] clarified the rela-

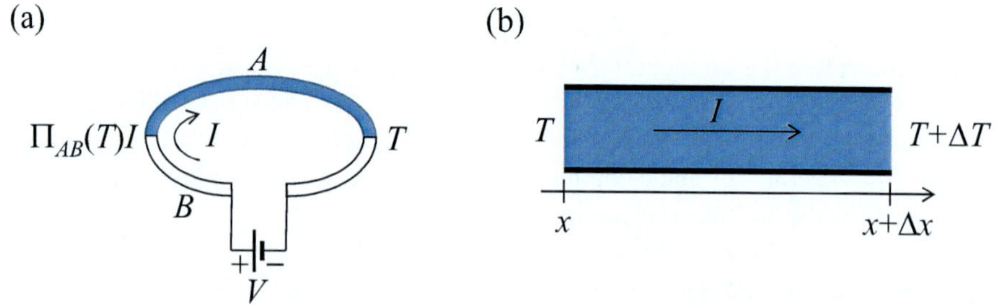

Fig. 7.2 Peltier effect and Thomson effect. **a** Thermocouple without temperature gradient. An electric current I is flowing. At the left (right) junction between the conductors A and B, the heat $\Pi_{AB}(T)I$ is emitted (absorbed) per unit time, i.e., $\Pi_{BA} = -\Pi_{AB}$. **b** Under the temperature gradient, an electric current is flowing in a conductor. To keep the temperature gradient, a heat emission/absorption occurs

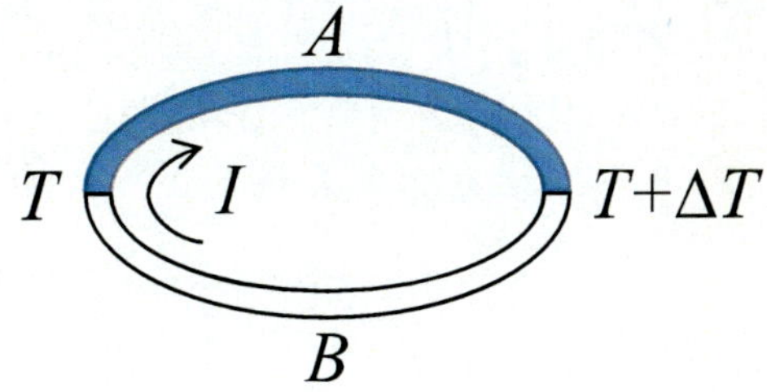

Fig. 7.3 Thermocouple with temperature gradient

tion between the effects. Let us discuss the relation on the thermodynamics consideration and this brings about the understanding of the thermopower Q clearly.

In Fig. 7.1a, because the conductor B is not singly connected one, the electromotive force (EMF) V_{AB} appears at the open ends of B. By closing this disconnected part, the EMF $V_{AB} = [Q_B(T) - Q_A(T)]\Delta T$ drives the electric current I (see Fig. 7.3).

Due to the Peltier effect, the heat $\Pi_{AB}(T)I$ ($\Pi_{AB}(T+\Delta T)I$) is emitted (absorbed) at the left (right) junction of the conductors A and B. The conductor A absorbs the heat Δq_A due to the Thomson effect,

$$\Delta q_a = \int_{x_l}^{x_r} \tau_A I \left(\frac{\partial T}{\partial x}\right) dx = \tau_A(T) I \Delta T \tag{7.7}$$

for small $\Delta T(>0)$, where τ_A is the Thomson coefficient of the conductor A, x is taken along A and left (right) end of A is x_l (x_r). In the same way, the conductor B emits the Thomson heat $\Delta q_B = \tau_B(T) I \Delta T$ with the Thomson coefficient of the conductor B, τ_B.

Because the total heat absorbed by the thermocouple is consumed by the electric power $IV_{AB} = I(Q_B - Q_A)\Delta T$, the first law of thermodynamics (the energy conservation law) is expressed by

$$I[Q_B(T) - Q_A(T)] = I\left[\tau_A(T) - \tau_B(T) + \Pi_{AB}(T+\Delta T) - \Pi_{AB}(T)\right]. \tag{7.8}$$

For $I \neq 0$, we find the first Kelvin relation,

$$(Q_B - Q_A) = \frac{d\Pi_{AB}}{dT} + \tau_A - \tau_B. \tag{7.9}$$

The heat emission/absorption by the Peltier and Thomson effect changes the sign by changing the electric current direction or the temperature gradient direction. On the other hand, the Joule heat does not depend on the signs of I and ΔT. This crucial difference is reasonably distinguished for the experimental setup with small enough ΔT and/or I. On this viewpoint, Thomson considered [34] the thermoelectric effects within a view of reversible-process-consideration. In this consideration, the net entropy production of the thermoelectric system shown in Fig. 7.3 is set to be zero and hence

$$0 = I\left[\int_T^{T+\Delta T} \frac{\tau_A(t) - \tau_B(t)}{t} dt + \frac{\Pi_{AB}(T + \Delta T)}{T + \Delta T} - \frac{\Pi_{AB}(T)}{T}\right]. \tag{7.10}$$

For $I \neq 0$ and $\Delta T \to 0$, we find

$$0 = \frac{\tau_A - \tau_B}{T} + \frac{1}{T}\frac{d\Pi_{AB}}{dT} - \frac{\Pi_{AB}(T)}{T^2}. \tag{7.11}$$

Equations (7.11) and (7.9) give the second Kelvin relation,

$$Q_B - Q_A = \frac{\Pi_{AB}(T)}{T}. \tag{7.12}$$

Note that, differentiation of Eq. (7.12) with respect to T leads to

$$\tau = T\frac{dQ}{dT}. \tag{7.13}$$

As discussed above, Thomson coefficient τ is the specific heat of electricity. Therefore, Eq. (7.13) tells us the meaning of the thermopower, i.e., the thermopower is the *entropy* of electricity.

The consideration of the thermoelectric effects on thermodynamics not only makes clear the meaning of the thermoelectric coefficients but also tells us the low-temperature behavior of them. Let us discuss the thermopower Q at the low-temperature $\Delta T \to 0$.

Figure 7.2a shows the thermoelectric system without temperature gradient and the electric current I is flowing in the thermocouple. Ohm's law tells us

$$I = \frac{V}{R_A + R_B}, \tag{7.14}$$

where R_A and R_B are the resistance of the conductors A and B, respectively. At the junctions between A and B, the heat $\Pi_{AB}(T)I$ is emitted/absorbed and hence the entropy change $\Delta s = \Pi_{AB}(T)I/T$ occurs at the junction(s). Equations (7.14) and (7.12) bring about

$$\Delta s = \frac{Q_B(T) - Q_A(T)}{R_A + R_B} V. \tag{7.15}$$

The third law of thermodynamics is expressed to be $\Delta s \to 0$ for $T \to 0$. This must be hold for any V and any thermocouple, and the dimension of the conductors is irrelevant for this consideration. Therefore, the third law of thermodynamics results in [35]

$$\lim_{T\to 0} Q/\rho = 0, \tag{7.16}$$

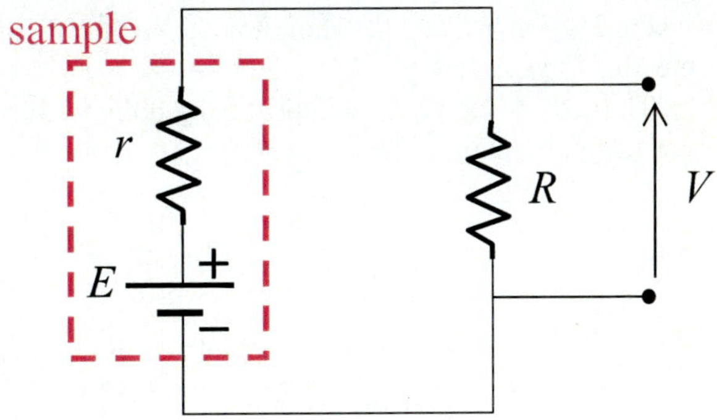

Fig. 7.4 A setup to measure electromotive force E of a sample

where ρ is the electrical resistivity. The conductors with $\lim_{T\to 0}(1/\rho) \neq 0$ must show $Q \to 0$ for $T \to 0$ within the relation Eq. (7.16) (even in the superconductors). The (intrinsic) semiconductors show $Q \sim 1/T$ at low temperatures, but the electrical resistivity does $(1/\rho) \sim \exp(-E_g/k_B T)$ $(E_g > 0)$ at the same time. Those are consistent with the third law of thermodynamics [35], i.e., Eq. (7.16).

Figure 7.4 shows an equivalent circuit of a setup to measure the electromotive force E of a sample. The sample itself has a resistance r, and hence $V = \frac{R}{r+R}E$. Therefore, for the sample with $\rho \sim \exp(E_g/k_B T)$ and $Q \sim 1/T$, the temperature dependence of V shows $1/[T \exp(E_g/k_B T)]$ and it gose to zero for $T \to 0$.

We see the low-temperature behavior of the thermopower by the entropy consideration. This consideration also tells us the high-temperature behavior of the thermopower. In the above discussion based on the thermodynamics, the detail of the electronic structure is not needed. In the next section, let us discuss the high-temperature thermopower with a help of statistical mechanics.

7.4 Thermopower: Entropy Carried by Electric Current

In the following sections, we discuss the high-temperature thermopower in some microscopic models. According to the statistical mechanics, the thermodynamic quantity is given by the statistical average: Within the grand-canonical ensemble formalism, more precisely, for a physical quantity expressed by an operator B, the expectation value $< B >$ is corresponding the thermodynamic quantity, i.e.,

$$< B >= \mathrm{Tr}(B\rho), \tag{7.17}$$

with

$$\rho = \frac{\exp\left[-\beta\left(H - \mu N\right)\right]}{\mathrm{Tr}\left\{\exp\left[-\beta\left(H - \mu N\right)\right]\right\}}, \tag{7.18}$$

where H, μ and N are the Hamiltonian, chemical potential, and number operator of electrons, respectively.

Thus, $\beta \to 0$ gives a canonical fixed point of the quantities in this formalism. In particular, by putting the relation

$$\beta = \frac{1}{k_B T} \tag{7.19}$$

(k_B: Boltzmann constant), the thermodynamic quantity at temperature T is discussed. Therefore, the fixed point for $\beta \to 0$ is sometimes called "high-temperature limit" consideration.

We can discuss the thermodynamic quantity by equation (7.18), but it is for equilibrium state. The thermopower Q, on the other hand, is a typical quantity in non-equilibrium state. Such response coefficient is discussed by Kubo formula. By the formula, the thermopower Q is expressed by the correlation functions of the electric current and energy current, and those are also expressed by equation (7.18). Even so, the microscopic expression of the thermopower is not simple and not easy to calculate, usually. However, the fixed point is given by rather simple consideration(s). Actually, it is shown [17, 36–38] that at $\beta \to 0$,

$$Q = \frac{\mu}{eT}, \tag{7.20}$$

with absolute value of an electron charge $e(> 0)$. Note that thermodynamics tells us

$$-\frac{\mu}{T} = \left.\frac{\partial s}{\partial N}\right|_{E,V}, \tag{7.21}$$

where s, N, E and V are entropy, number of particles, internal energy, and volume of the system, respectively, and hence the expression Eq. (7.20) relates the entropy per carrier and thermopower Q. Let us discuss this fixed point Eq. (7.20).

7.5 High-Temperature Thermopower of Single Band Hubbard Model

We start with a single band Hubbard model,

$$H = -t \sum_{<ij>,\sigma} \left(c^\dagger_{i,\sigma} c_{j,\sigma} + h.c. \right) + U \sum_i n_{i\uparrow} n_{i\downarrow}, \tag{7.22}$$

with the hopping integral t of an electron and the Coulomb interaction U, and other notations are standard. In this model, we can discuss two high-temperature limits [17, 36], i.e., (i) t, $U < k_B T$ and $k_B T \to \infty$ and (ii) $k_B T \to \infty$ with $t < k_B T < U$. For

Fig. 7.5 Schematic representation of electron occupation at a site. **a** No electron on the site. **b** An up-spin electron on the site. **c** A down-spin electron on the site. **c** The doubly occupied state

these cases (i) and (ii), we find the high-temperature thermopower Eq. (7.20) by the expectation value of electron number n_e, i.e.,

$$n_e = \frac{\mathrm{Tr}\,\{N \exp\left[-(H-\mu N)/\left(k_B T\right)\right]\}}{\mathrm{Tr}\,\{\exp\left[-(H-\mu N)/\left(k_B T\right)\right]\}}. \tag{7.23}$$

Let us first examine the high-temperature thermopower in Case (i).

Case (i)
In this case, $t/\left(k_B T\right)$ and $U/\left(k_B T\right)$ go to zero, and hence Eq. (7.23) becomes

$$n_e = \frac{0+2\exp\left[\mu/\left(k_B T\right)\right]+2\exp\left[2\mu/\left(k_B T\right)\right]}{1+2\exp\left[\mu/\left(k_B T\right)\right]+\exp\left[2\mu/\left(k_B T\right)\right]} = \frac{2\exp\left[\mu/\left(k_B T\right)\right]}{1+\exp\left[\mu/\left(k_B T\right)\right]}. \tag{7.24}$$

As a result, we find

$$Q_{(\mathrm{i})} = -\frac{k_B}{e}\ln\left(\frac{2-n_e}{n_e}\right). \tag{7.25}$$

Note that, all the states at a site (see Fig. 7.5) contribute to derivation of Eq. (7.24). In the case (ii), on the other hand, we see that the effect of Coulomb interaction plays an important role for the high-temperature thermopower.
Case (ii)
In this case, the high-temperature limit is taken under the condition $k_B T < U$. In other words, for the calculation Eq. (7.23), the strong U limit, $U/\left(k_B T\right) \to \infty$, is taken first, and later the high-temperature limit is considered. In such strong U system, the electronic structure is described by the "Hubbard band" consideration. Here let us suppose the chemical potential μ is lying in the lower-Hubbard-band. Because $\exp\left[-U/\left(k_B T\right)\right] \to 0$ in the present case, the contribution from the doubly occupied state (see Fig. 7.5d) is suppressed, and (7.23) gives

$$n_e = \frac{0+2\exp\left[\mu/\left(k_B T\right)\right]+2\times 0}{1+2\exp\left[\mu/\left(k_B T\right)\right]+0}. \tag{7.26}$$

This results in,

$$Q_{(\mathrm{ii})} = -\frac{k_B}{e}\ln\left(2\frac{1-n_e}{n_e}\right) = -\frac{k_B}{e}\ln\frac{2x}{1-x} = -\frac{k_B}{e}\ln 2 - \frac{k_B}{e}\ln\frac{x}{1-x} \tag{7.27}$$

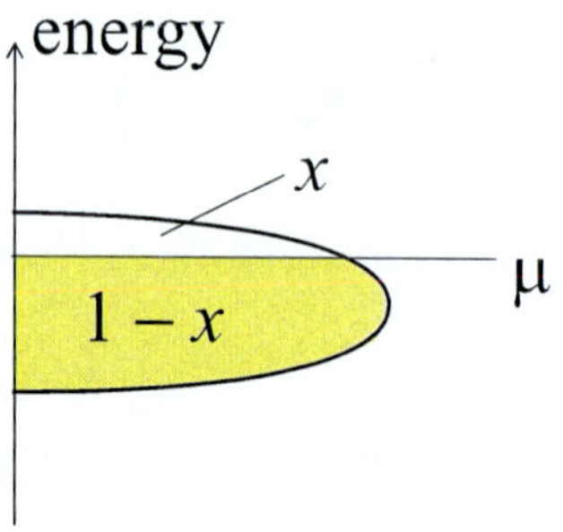

Fig. 7.6 Schematic representation of electronic state of free-electron system. The area of density-of-states (DOS) below (above) the chemical potential μ is $1 - x$ (x)

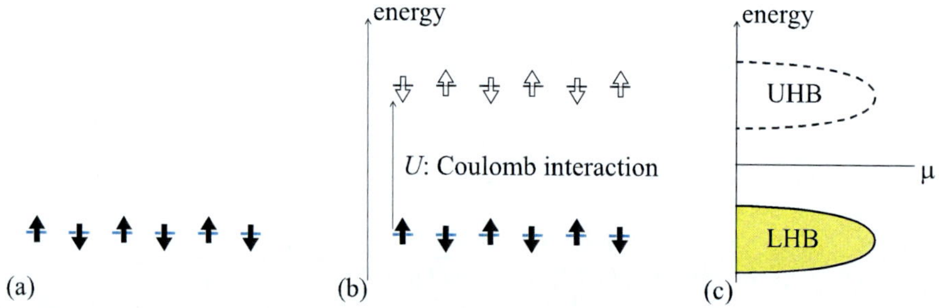

Fig. 7.7 Schematic representation of Mott-insulating-state. **a** Real-space representation of the Mott-insulating-state. Each site is occupied by an electron. The electron spin is expressed by arrow. **b** The *effective* energy level for the additional electron: For each site, it costs an energy U to add an electron with opposite spin (indicated by open arrow) to the occupied electron's spin. **c** In the half-filled electronic state, the singly occupied states form the lower Hubbard band (LHB) and the DOS is totally occupied. The upper Hubbard band (UHB) corresponding to that by the doubly occupied states is empty. The chemical potential μ is located at the center between LHB and UHB

where $x = 1 - n_e$ represents the *hole* number. Note that the first term of the final expression in (7.27), ln 2 comes fron the spin degeneracy of the singly occupied state(s) (see Fig. 7.5b and c).

The last term of the end of Eq. (7.27), $Q = -\frac{k_B}{e} \ln \frac{x}{1-x}$ is called Heikes formula which is used to discuss the thermopower of the semiconductors. In many cases, the electronic state of the semiconductors is discussed by spinless electronic state with narrow bandwidth (see Fig. 7.6): In this case, the chemical potential μ is lying on the energy band and electron (hole) concentration is $1 - x$ (x). The ratio between numbers of the vacant and occupied states, $x/(1 - x)$ appears in the Heikes formula.[1]

We find interesting similarity/difference between the Heikes formula and Eq. (7.27) which reflects the electronic state of the doped Mott insulator [17, 39–41]: Fig. 7.7 shows the schematic representation of the Mott-insulating state, i.e., the half-filled

[1] In the non-interacting spinful electron system, the spin degeneracy is canceled out for the ratio between the numbers of the vacant and occupied states. Therefore, the role of spin degeneracy is often neglected in the semiconductor physics.

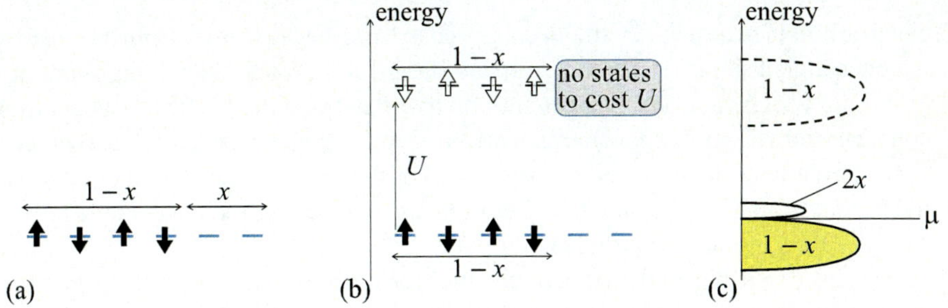

Fig. 7.8 Schematic representation of electronic state of doped Mott insulator. **a** Holes with an amount of x are introduced to the Mott-insulating state. Therefore, the ratio between unoccupied and occupied site is $x/(1-x)$. **b** For the occupied site, it costs an energy U to add an electron with opposite spin (indicated by open arrow) to the occupied electron's spin, but no energy costs to add an electron to the unoccupied site. **c** Schematic representation of the DOS for the doped Mott insulator. By hole-doping, the chemical potential μ is lying on the LHB and total states below μ are $1-x$, i.e., the amount of total occupied sites. The DOS of the UHB appears due to the energy cost by the Coulomb interaction U at the occupied sites. Therefore, the amount of total states of the UHB is expressed to be $1-x$. At the unoccupied site, an electron is able to be added without U but the electron has the spin degeneracy, i.e., spin-up or spin-down. As a result, the amount of *vacant* states of the LHB is $2x$ being (hole concentration)×(spin degeneracy)

electronic state. To simplify the discussion, we suppose the case for $U/t \to \infty$ and the effect of t is neglected. Figure 7.7a represents the real-space representation of the Mott-insulating state. Each site is occupied by an electron. The energy gap E_{gap} of the insulating state for the N electron system is defined by

$$E_{gap} = [E_{N+1} - E_N] + [E_{N-1} - E_N] = E_{N+1} + E_{N-1} - 2E_N, \qquad (7.28)$$

where E_{N+1}, E_N and E_{N-1} are the total energy of the system with $N+1$, N, and $N-1$ electrons. In the present case, $E_{gap} = U$. The change in energy $E_{N+1} - E_N$ $(E_{N-1} - E_N)$ is corresponding to the procedure to add (remove) an electron to (from) the system, and total number of ways of the procedure determines the total weight of DOS of the upper (lower) Hubbard band (see Fig. 7.7b). Because of the strong U, the doubly occupied state (see Fig. 7.5d) is suppressed. Thus, the lower Hubbard band (LHB) (upper Hubbard band (UHB)) is totally occupied (empty) and the chemical potential μ stays at the center between UHB and LHB (see Fig. 7.7c). In this insulating state, the (empty) DOS of UHB is due to the strong U of the *occupied* site (see Fig. 7.5b and c).

Figure 7.8 shows the schematic representation of the hole-doping on the Mott insulator. By introducing holes with an amount of x to the Mott-insulating state, the electronic states will be expressed by the occupied and unoccupied sites by electrons as shown in Fig. 7.8a, b. The amount of the occupied site contributes to the electron occupancy below μ, so that it is $1-x$ as seen in Fig. 7.8b. To add an electron at the occupied site, the situation is similar to the case discussed above using Fig. 7.7. Therefore, the amount of total states of the UHB is $1-x$. At the unoccupied site, an electron is able to be added without energy cost. Here, because of the spin degeneracy, the unoccupied site is responsible for *two* states for the vacant low lying states at μ (see Fig. 7.8b). As a result, the amount of *vacant* states of the LHB is $2x$ being (hole concentration)$\times$(spin degeneracy). This is a typical consequence of the *Mottness* of the strongly correlated electron systems. For the high-temperature thermopower, the Mottness appears, i.e., the ratio between areas of the vacant and occupied states, $2x/(1-x)$ is found in Eq. (7.27).

We discussed the high-temperature thermopower of the single band Hubbard model. Similar to this, we can discuss the case of multi-band Hubbard model. In the next section, let us discuss the three-band case to discuss the thermopower for the $3dt_{2g}$ electrons.

7.6 High-Temperature Thermopower of Three-Band Hubbard Model

The three-band Hubbard model will be given by

$$H = -\sum_{<il,jk>,\sigma} t_{il,jk}\left(c^{\dagger}_{il,\sigma}c_{jk,\sigma} + h.c.\right) + U\sum_{il} n_{il\uparrow}n_{il\downarrow} + U'\sum_{i,l\neq k} n_{il}n_{ik} + J\sum_{i,l\neq k,\sigma\sigma'} c^{\dagger}_{il\sigma}c^{\dagger}_{ik\sigma'}c_{il\sigma'}c_{ik\sigma}, \quad (7.29)$$

with the hopping integrals $t_{il,jk}$ of an electron and the Coulomb interactions U, U' and J, where l and k $(= 1 \sim 3)$ are the indices of the orbitals at the sites i and j, respectively, and $n_{il} = n_{il\uparrow} + n_{il\downarrow}$.

In the present case, the two high-temperature limits [29] corresponding to the cases (i) and (ii) discussed in previous section will be

(i)' $t_{il,jk} < k_BT \to \infty$ and k_BT is larger than the Coulomb interactions, and

(ii)' $t_{il,jk} < k_BT \to \infty$ but the Coulomb interactions are larger than k_BT, respectively. In the high-temperature consideration, the energy difference of the three t_{2g} levels are neglected.

Case (i)'

In this case, all the states at a site i corresponding to d^m state with $m = \sum_{l=1}^{3} n_{il} = 0 \sim 6$ contribute to the calculation, so that Eq. (7.23) gives

$$n_e = \frac{\sum_{m=0}^{6}\binom{6}{m} \times m \times \exp\left[m\mu/(k_BT)\right]}{\sum_{m=0}^{6}\binom{6}{m} \times \exp\left[m\mu/(k_BT)\right]} = \frac{6\exp\left[\mu/(k_BT)\right]}{1+\exp\left[\mu/(k_BT)\right]}. \quad (7.30)$$

Therefore,

$$Q_{(\mathrm{i})'} = -\frac{k_B}{e} \ln\left(\frac{6 - n_e}{n_e}\right). \tag{7.31}$$

Case (ii)'
In the strong Coulomb interaction case, the Hubbard band consideration is important. In case (ii) discussed in the previous section, the chemical potential μ is lying in the lower-Hubbard-band. In particular, the states with no-electron site (Fig. 7.5a) and singly occupied site (Fig. 7.5b and c) contribute to Eq. (7.23). Those electronic states are corresponding to d^0 and d^1 states in view of electron occupation. For the three-band Hubbard model Eq. (7.29), the chemical potential μ specifies a Hubbard band formed by the d^{m-1} and d^m states, and those states contribute to Eq. (7.23).

Equation (7.23) becomes

$$n_e = \frac{(m-1)g_h + m g_e \exp\left[\mu/\left(k_B T\right)\right]}{g_h + g_e \exp\left[\mu/\left(k_B T\right)\right]}, \tag{7.32}$$

where g_h and g_e are the *local* spin/orbital degeneracy of the d^{m-1} and d^m states. We find

$$Q_{(\mathrm{ii})'} = -\frac{k_B}{e} \ln\left[\frac{g_e}{g_h} \frac{m - n_e}{1 - (m - n_e)}\right] = -\frac{k_B}{e} \ln\frac{g_e}{g_h} - \frac{k_B}{e} \ln\frac{x}{1-x}, \tag{7.33}$$

where $x = m - n_e$ represents the hole number for the present case. Note that, Eq. (7.33) is reduced to Eq. (7.27) for $\{m = 1, g_e = 2, g_h = 1\}$.

7.7 Thermopower of $3dt_{2g}$ Electron Systems: Cobalt Oxide and Vanadium Oxide

In this section, we discuss the "high-temperature" behaviors of the thermopower in $3dt_{2g}$ electron systems. Figure 7.9 shows the thermoelectric response of the layered cobalt oxide Na_2CoO_4 [1] (In Ref. [1], S is used for the thermopower.). This cobalt oxide is a metallic system. Therefore, as discussed in Sect. 7.3, the thermopower goes to zero for $T \to 0$. With increasing temperature, the thermopower is positive and increasing. At room temperatures, the thermopower reaches $\sim 100\,\mu$V/K.

Figure 7.10 shows the temperature dependence of the thermopower of the vanadium oxide $La_{1-x}Sr_xVO_3$ [29], as another example of the thermoelectric response in $3dt_{2g}$ electron system. In this case, the thermopower at high temperatures shows *nonmonotonic* temperature dependence: The thermopower for $x = 0.18$ has a peak structure at $T \sim 700$K. The peak height is decreasing with increasing x, and finally the sign of the thermopower is negative for whole temperature region in this measurement.

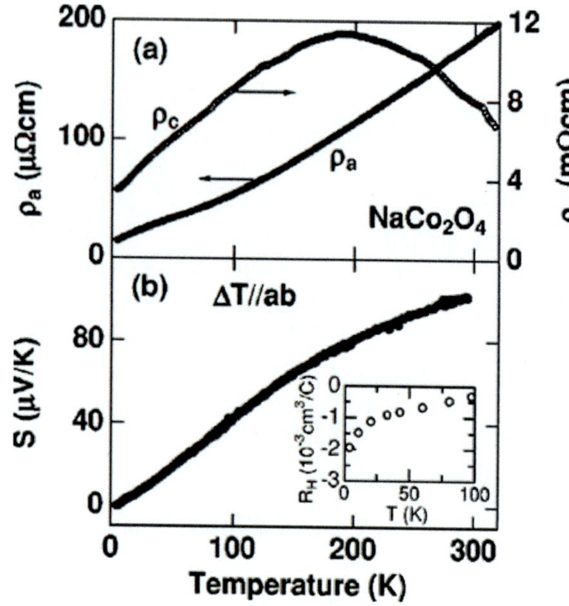

Fig. 7.9 Thermoelectric response of the layered cobalt oxide Na_2CoO_4 shown in Ref. [1]

Let us apply the considerations (i)' and (ii)' discussed in the previous Sect. 7.6, to the high-temperature thermopower in the cobalt oxide and the vanadium oxide.

When we use the chemical formula of the cobalt oxide Na_2CoO_4 to evaluate the valence of the cobalt ion, it gives +3.5. Figure 7.11a shows the electron occupation of the $3d_{2g}$ orbitals for Co^{3+} state, i.e., d^6 one. Therefore, $n_e = 5.5$ is corresponding to the average valence of the cobalt ion 3.5+. In the case (i)', $n_e = 5.5$ and Eq. (7.31) gives $Q_{(\mathrm{i})'} = 206\mu$V/K. In the case (ii)', the strong Coulomb interaction works, and hence the average valence of the cobalt ion 3.5+ corresponds to that the ratio of Co^{3+} to Co^{4+} is unity. The electronic state of Co^{3+} (d^6 state) shown in Fig. 7.11a is uniquely determined. On the other hand, for the d^5 state of Co^{4+} ion, there are six ways to fill the $3dt_{2g}$ orbitals by five electrons and Fig. 7.11b is one of the electron configuration. Consequently, the case (ii)' with $\{m = 6, n_e = 5.5, g_e = 1, g_h = 6\}$ gives $Q_{(\mathrm{ii})'} = 154\mu$V/K. In the cobalt oxide Na_2CoO_4, the positive and large $Q_{(\mathrm{ii})'}$ and $Q_{(\mathrm{i})'}$ are consistent with the experiment.

In the case of the vanadium oxide $La_{1-x}Sr_xVO_3$, the valence of vanadium ion is $+3 + x$ and $n_e = 2 - x$ is corresponding to the electron occupation of the $3dt_{2g}$ orbitals. Consequently, Eq. (7.31) gives

$$Q_{(\mathrm{i})'} = -\frac{k_B}{e} \ln\left(\frac{4+x}{2-x}\right). \tag{7.34}$$

On the other hand, in the strong Coulomb interaction(s) regime, i.e., for the case (ii)', the $3dt_{2g}$ electron configuration of the vanadium ions will be expressed by d^2 and d^1 states as shown in Fig. 7.12. In particular, the total-spin magnitude S of the d^2 state takes $S = 1$ in this case, so that the g_e =(spin degeneracy)×(orbital

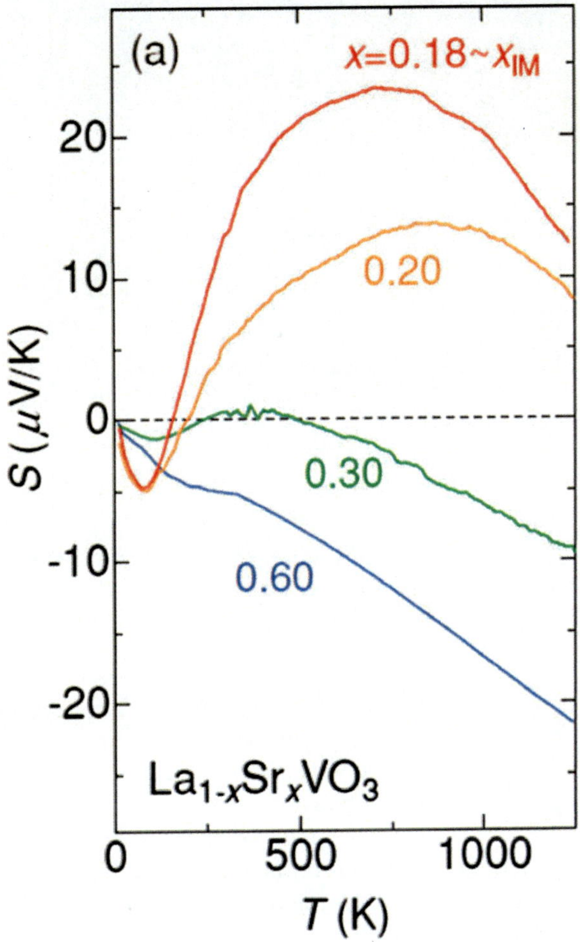

Fig. 7.10 Thermoelectric response of the vanadium oxide $La_{1-x}Sr_xVO_3$ shown in Ref. [29]

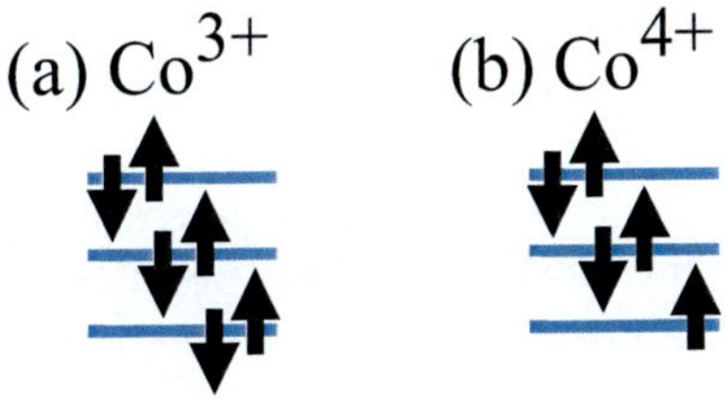

Fig. 7.11 Schematic representation of the electron occupation of cobalt ions **a** Co^{3+} and **b** Co^{4+} in $3dt_{2g}$ orbitals

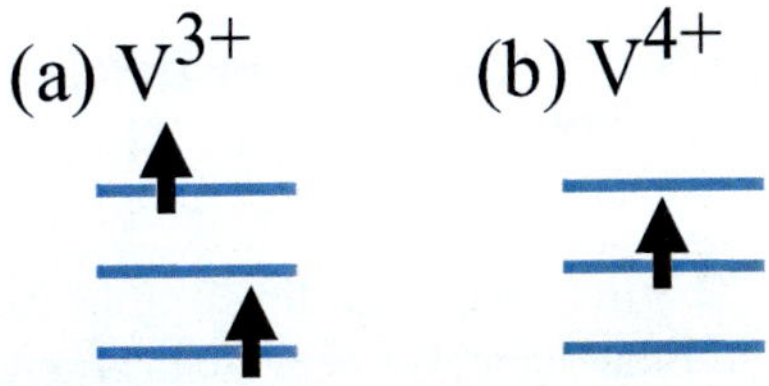

Fig. 7.12 Schematic representation of the electron occupation of vanadium ions **a** V^{3+} and **b** V^{4+} in $3dt_{2g}$ orbitals

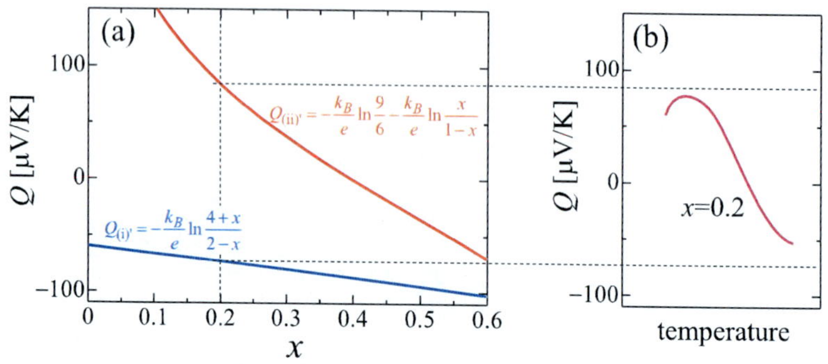

Fig. 7.13 The high-temperature thermopower $Q_{(i)'}$ and $Q_{(ii)'}$ for the vanadium oxide $La_{1-x}Sr_xVO_3$ (see text). **a** Doping x dependence. **b** Expected temperature dependence of Q for $x = 0.2$ at high temperatures

degeneracy)$= 3 \times 3 = 9$. In the same way, the local degeneracy of the d^1 state is $g_h = 6$. Therefore, for $\{m = 2, g_e = 9, g_h = 6\}$, we find

$$Q_{(ii)'} = -\frac{k_B}{e}\ln\frac{9}{4} - \frac{k_B}{e}\ln\frac{x}{1-x}. \quad (7.35)$$

Figure 7.13a shows the hole concentration x dependence of $Q_{(i)'}$ and $Q_{(ii)'}$ for the present case. From this viewpoint, let us discuss the temperature dependence of the thermopower of the vanadium oxide $La_{1-x}Sr_xVO_3$ in high-temperature region, i.e., from $\sim$ 1000K down to room temperatures.

In reality, the Coulomb interaction is finite, so that the $final$ high-temperature limit of the thermopower should be given by $Q_{(i)'}$. From the high-temperature side, with decreasing temperature, the effect of Coulomb interactions will show up, and hence the thermopower will approach the value given by $Q_{(ii)'}$. Figure 7.13b shows the expected temperature dependence of Q for $x = 0.2$ at high temperatures. As seen in Fig. 7.13, $Q_{(ii)'}$ is always larger than $Q_{(i)'}$. This is consistent with the temperature dependence of the thermopower of the $La_{1-x}Sr_xVO_3$ at the high-temperature region (see Fig. 7.10), i.e., the thermopower goes up with decreasing temperature from the high-temperature end of this measurement. In particular, for $x \sim 0.3$, the thermopower becomes negative for whole temperature region. This is also consistent with the high-temperature consideration of the thermopower shown in Fig. 7.13.

7.8 Summary and Discussions: Role of Spin and Orbital Degeneracy on Thermopower

Thermopower is a direct probe of entropy flow by electric current. The spin and orbital degrees of freedom carried by electrons can contribute to the entropy flow. However, such degrees are not always activated: For spin degree, strong magnetic anisotropy and/or magnetic ordering will suppress the entropy flow by spin(s). For

Table 7.1 Thermopower by spin and orbital degrees of freedom

$Ti^{3+}(d^1)$, $Ti^{4+}(d^0)$	6/1	$-154\mu V/K$
$V^{3+}(d^2)$, $V^{4+}(d^1)$	9/6	$-35\mu V/K$
$Mn^{3+}(d^4)$, $Mn^{4+}(d^3)$	10/4	$-79\mu V/K$
$Co^{3+}(d^6)$, $Co^{4+}(d^5)$	1/6	$+154\mu V/K$

orbital degree, the Jahn-Teller effect is the candidate to suppress the entropy flow by orbital(s). These suppression effects will be enhanced in the systems with bipartite lattice structure by spin/orbital ordering. The effect of the Fermi degeneracy is also the candidate to suppress the entropy flow by spin and orbitals. In particular, for good metals which electronic states are well described by free electrons, the electron transport is only determined by the electrons at (around) the Fermi surface, and hence the shape of DOS is the key to understand the thermopower. At low temperatures, by the effect of Fermi degeneracy, the contribution from the spin and orbital degrees of freedom will be suppressed, even in the strongly correlated electron systems.

Strong correlation disturbs one-electron picture in the electron transport, i.e., the lifetime or coherence of the electron is limited. A direct probe of the coherence of the electron transport is given by angle-resolved photoelectron spectroscopy (ARPES) measurement. In the cobalt oxide Na_2CoO_4, it has been shown [12] that the coherence is observed at well below room temperatures but it is lost at $\sim$ 200K. At this temperature, the observed thermopower of the cobalt oxide Na_2CoO_4 reaches $\sim 80\mu V/K$. At $\sim$ 200K, the thermopower of the vanadium oxide $La_{1-x}Sr_xVO_3$ with $(x > 0.3)$ shows the increasing behavior with increasing temperature. Note that the width of the coherence peak determines a characteristic energy scale of the electron transport properties. At the temperature beyond the characteristic energy scale, it seems that the *high-temperature* is achieved, and hence the formula Eq. (7.33) works where the coherence is lost.

This formula Eq. (7.33) tells us a new path to the thermoelectric material by doped Mott insulators: In the formula Eq. (7.33), the contribution from charge, spin, and orbital degrees of freedom to the thermopower is transparent, i.e., the last term (Heikes formula) $-\frac{k_B}{e}\ln\frac{x}{1-x}$ represents the *charge* entropy contribution and the first term $-\frac{k_B}{e}\ln\frac{g_e}{g_h}$ comes from the entropy flow by *spin* and *orbital*. In the conventional thermoelectrics which rely only on the charge degree of freedom, the thermopower is enhanced by decreasing carrier concentration, but at the same time, the resistivity also goes up. This is a drag on development. The strong Coulomb interaction activates the entropy flow by *spin* and *orbital* and this new channel to the thermopower is essentially different from charge one. This gives a chance to develop the material with low resistivity by large charge-carrier density and high-thermopower by the spin-orbital entropy flow. The table 7.1 summarizes the thermopower by spin and orbital degrees of freedom with sufficiently strong Coulomb interaction effect.

Acknowledgements We would like to thank K. Tsutsui, S. Maekawa, S. Okamoto, M. Mori, and M. Matsuo for helpful discussions. This paper is dedicated to the memory of Professor Sumio Ishihara, who recently passed away, for his kind hospitality and valuable discussions.

References

1. I. Terasaki, Y. Sasago, K. Uchinokura, Phys. Rev. B **56**, R12685 (1997)
2. D.J. Singh, Phys. Rev. B **61**, 13397 (2000)
3. T. Itoh, I. Terasaki, Jpn. J. Appl. Phys. **39**, 6658 (2000)
4. A.C. Masset et al., Phys. Rev. B **62**, 166 (2000)
5. R. Funahashi et al., Jpn. J. Appl. Phys. **39**, L1127 (2000)
6. W. Koshibae, K. Tsutsui, S. Maekawa, Phys. Rev. B **62**, 6869 (2000)
7. T. Motohashi et al., Appl. Phys. Lett. **79**, 1480 (2001)
8. K. Fujita, T. Mochida, K. Nakamura, Jpn. J. Appl. Phys. **40**, 4644 (2001)
9. S. Hébert et al., Phys. Rev. B **64**, 172101 (2001)
10. Y. Miyazaki et al., J. Phys. Soc. Jpn. **71**, 491 (2002)
11. I. Terasaki, I. Tsukada, Y. Iguchi, Phys. Rev. B **65**, 195106 (2002)
12. T. Valla et al., Nature **417**, 627 (2002)
13. Y. Wang et al., Nature **423**, 425 (2003)
14. T. Takeuchi, Phys. Rev. B **69**, 125410 (2004)
15. M.L. Foo, Phys. Rev. Lett. **92**, 247001 (2004)
16. M.Z. Hasan, Phys. Rev. Lett. **92**, 246402 (2004)
17. S. Maekawa et al., *Physics of Transition Metal Oxide, Springer Series in Solid-State Sciences*, vol. 144 (Springer-Verlag, Berlin Heidelberg, 2004)
18. T. Okuda et al., Phys. Rev. B **72**, 144403 (2005)
19. S. Okada, I. Terasaki, Jpn. J. Appl. Phys. **44**, 1834 (2005)
20. Y. Okamoto et al., J. Phys. Soc. Jpn. **75**, 023704 (2006)
21. M. Lee et al., Nature Materials **5**, 537 (2006)
22. V.S. Oudovenko et al., Phys. Rev. B **73**, 035120 (2006)
23. Y. Ono et al., Jpn. J. Appl. Phys. **46**, 1071 (2007)
24. G.B. Wilson-Short et al., Phys. Rev. B **75**, 035121 (2007)
25. Y. Ishida et al., J. Phys. Soc. Jpn. **76**, 103709 (2007)
26. K. Kuroki, R. Arita, J. Phys. Soc. Jpn. **76**, 083707 (2007)
27. T. Motohashi et al., Phys. Rev. B **83**, 195128 (2011)
28. W. Xu, C. Weber, G. Kotliar, Phys. Rev. B **84**, 035114 (2011)
29. M. Uchida et al., Phys. Rev. B **83**, 165127 (2011)
30. M. Matsuo et al., Phys. Rev. B **84**, 153107 (2011)
31. J. Hejtmánek, Z. Jirák, J. Šebek, Phys. Rev. B **92**, 125106 (2015)
32. I. Terasaki, APL Materials **4**, 104501 (2016)
33. N.W. Ashcroft, N.D. Mermin, *Solid State Physics* (Thomson Learning, U. K., 1976)
34. W. Sir, *Thomson, Mathematical and Physical papers*, vol. 1 (Cambridge University Press, Cambridge, 1882)
35. M. Kohler, Abhandlungen der Braunschweigischen Wissenschaftlichen Gesellschaft **3**, 49 (1951)
36. P.M. Chaikin, G. Beni, Phys. Rev. B **13**, 647 (1976)
37. A. Oguri, S. Maekawa, Phys. Rev. B **41**, 6977 (1990)
38. G. Pálsson, G. Kotliar, Phys. Rev. Lett. **80**, 4775 (1998)
39. H. Eskes, M.B.J. Meinders, G.A. Sawatzky, Phys. Rev. Lett. **67**, 1035 (1991)
40. Y. Ohta et al., Phys. Rev. B **46**, 14022 (1992)
41. M.B.J. Meinders, H. Eskes, G.A. Sawatzky, Phys. Rev. B **48**, 3916 (1993)

Index

Y. Okimoto et al. (eds.), *Spin-Crossover Cobaltite*, Springer Series in Materials Science 305, https://doi.org/10.1007/978-981-15-7929-5

O

P

R

S

T

U